MW01258797

WONDERS OF
CREATION

DESIGN IN A FALLEN WORLD

STUART BURGESS AND ANDY MCINTOSH EDITOR: BRIAN EDWARDS

A Division of New Leaf Publishing Group
www.masterbooks.com

First printing Master Books® edition: December 2018
Originally published by Day One Publications (2017)

Master Books®, P.O. Box 726, Green Forest, AR 72638
Master Books® is a division of the New Leaf Publishing Group, Inc.

ISBN: 978-1-68344-166-3
ISBN: 978-1-61458-695-1 (digital)

Design by Steve Devane

Please consider requesting that a copy of this volume be purchased by your local library system.

Printed in China

Please visit our website for other great titles:
www.masterbooks.com

For information regarding author interviews,
please contact the publicity department at (870) 438-5288.

Dedication

Dedicated to the memory of Professor Verna Wright (1928–1998) who, as well as being a world renowned scientist in medicine (rheumatology), also clearly believed in biblical creation. Well known for his outspoken Christian stand and his masterful lectures against evolution, Verna Wright greatly helped and encouraged one of the authors (ACM) of this current book in developing scientific arguments from his own scientific discipline of engineering and mathematics.

WONDERS OF CREATION

DESIGN IN A FALLEN WORLD

The Wisdom of God

Beyond the Book of Genesis, one of the most beautiful passages in the Bible describes, in the symbol of a person, the wisdom of the God who made everything.

"The LORD brought me forth as the first of his works,
before his deeds of old;
I was appointed from eternity,
from the beginning, before the world began.
When there were no oceans, I was given birth,
when there were no springs abounding with water;
before the mountains were settled in place,
before the hills, I was given birth,
before he made the earth or its fields
or any of the dust of the world.
I was there when he set the heavens in place,
when he marked out the horizon on the face of the deep,
when he established the clouds above
and fixed securely the fountains of the deep,
when he gave the sea its boundary
so that the waters would not overstep his command,
and when he marked out the foundations of the earth.
Then I was the craftsman at his side.
I was filled with delight day after day,
rejoicing always in his presence,
rejoicing in his whole world
and delighting in mankind."

Proverbs 8:22–31 (NIV)

01 Since the Creation of the World

"Since the creation of the world God's invisible qualities — his eternal power and divine nature — have been clearly seen, being understood from what has been made" (Romans 1.20, NIV).

Everyone agrees that if the universe came into existence "by chance," it did so against immeasurable odds. It follows that if every living thing evolved step by step over millions of years into the intricate, complex, and kaleidoscopic beauty and order that we see around us, this also was against these incalculable odds. We may go further and claim that if the vital interrelationships of the universe evolved — the precise movements of the planets, the regular seasons of the year, the reliance of all living things upon each other — they did so against the same unimaginable odds.

In other words, we must conclude either that everything — each individual detail — in the known universe, and especially planet Earth, evolved against a series of unimaginable and unbelievable odds — or we should look for an alternative and more reasonable explanation.

It is that more reasonable explanation which is presented in this book.

No human being was there at the start so, left to ourselves, we cannot know for certain how it all began. As one popular scientist frankly admitted, "There is no scientific mechanism to explain how the universe began."[1] In other words, the honest scientist must concede that our current views about the origin of everything are often speculation — guesswork based upon the little that we actually know for certain and what we think might have been.

1 Prof. Brian Cox in his BBC *Human Planet* series, October 2014.

Continually representing this as fact does not turn speculation into reality.

In the pages that follow, you will experience a few of the incredible complexities of the world around us as you understand the things that are made. You will see, over and over, the meaning of "irreducible complexity" — that so much in creation is so complex that it must be complete at once to work at all. A rational mind will face the challenge of whether this is all the result of an unlimited number of immeasurable odds, or whether a more satisfying and reasonable explanation is that the invisible character and the eternal power of a Creator is clearly seen.

Understanding origins is much more important than conjecture and guesswork by some parts of the scientific community. If we know how everything began, that might give us an explanation of how it will all end, and equally important, how we explain where we are now. The knowledge of origin and destination will tell us much about the route between the two.

Sadly, that route is plainly not all that it might be. The beauty and order, the incredible harmony and diversity, is too often shattered by violence and pain, disorder and death. Before our journey in this book concludes, we will face this issue and the question posed by theologians, scientists, and philosophers for millennia: "What is the purpose of it all?"

But first, enjoy the panorama of a creation so beautifully detailed, ordered, and complex that it would be unbelievable if it were not there in front of us.

If we know how everything began, that might give us an explanation of how it will all end, and equally important, how we explain where we are now.

02 Land Mammals — Horses

The horse has a super-sized heart and lungs that enable it to be fast, strong, and powerful. It is ideal for humans because it can be ridden, trained, and bred into many varieties and for many uses.

1 Wild horses in Patagonia, Chile

Most horses in modern times are domesticated and are used for work, sport, or leisure. However, there are still wild horses in many parts of the world. Figure 1 shows wild horses in the National Park of Torres del Paine, Patagonia, Chile. Large horses can be as tall as 18 hands (6 feet) at the "withers," which is the ridge between the shoulder blades of the horse. Ponies are small horses, usually defined as horses that measure less than 14.2 hands (4.8 feet) when fully grown.

Lungs and heart for power

The fastest horses can run at speeds of up to 50 mph, which is around double the speed of the best human athletes. One of the reasons horses can run so fast is their huge lungs and heart and consequent capacity to extract large quantities of oxygen for energy. Figure 2 shows how horses have enormous lungs compared to humans.

The whole cardiovascular system of horses is designed for power and speed. The lungs are so big that they can take in 15 l of air per breath, and an incredible 1,800 l of air intake per minute compared with 150 l per minute in humans. Horses also have a large heart and a large quantity of blood of around 50 l. This volume of blood enables vast amounts of oxygen to be carried around the body to drive the muscles. Even the action of running is designed to help automatically open and close the lungs.

Car designers commonly increase the power in a car by using a turbocharger to compress air and thus

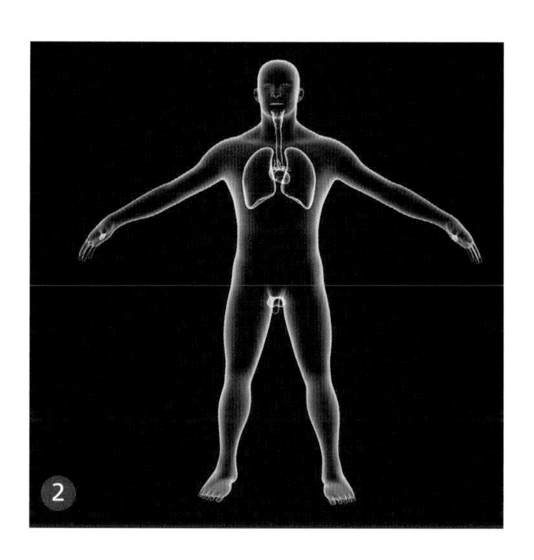

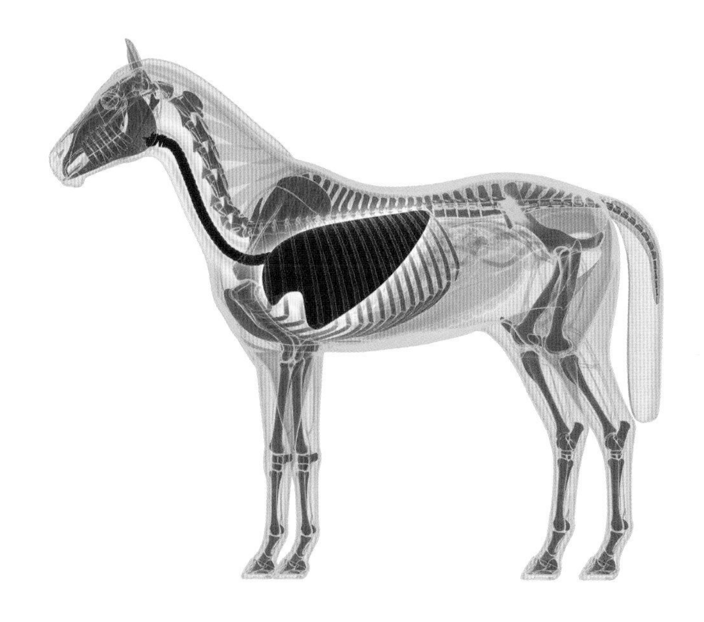

2 The lung capacity of horses is around 10 times larger than humans.

3 Horse legs have a suspension system to avoid shock loads.

get more oxygen into the engine. In the same way that turbochargers require expert design, so does the cardiovascular system of the horse.

Designer suspension

One of the challenges of running and jumping is that a strong suspension system is needed in the front and rear legs of the horse to reduce shock loads and give stability. The rear legs have joints that are bent at just the right angle when standing so that any shock load immediately bends the joints rather than sending the shock load up through the legs.

Even though the front legs of horses are straight, they also have a special feature to reduce shock loads. The front legs do not directly attach to the spine (there is no collar bone) but are suspended in place by muscles and tendons. This means that the body is elastically slung between two pillars resulting in a soft suspension system. This is particularly important for jumping because horses land heavily on their front two legs as seen in figure 3. When a racing horse is galloping, its weight may be taken by just one or two legs at a time and in this case a single leg may have to support over 1,000 lbs. in weight.

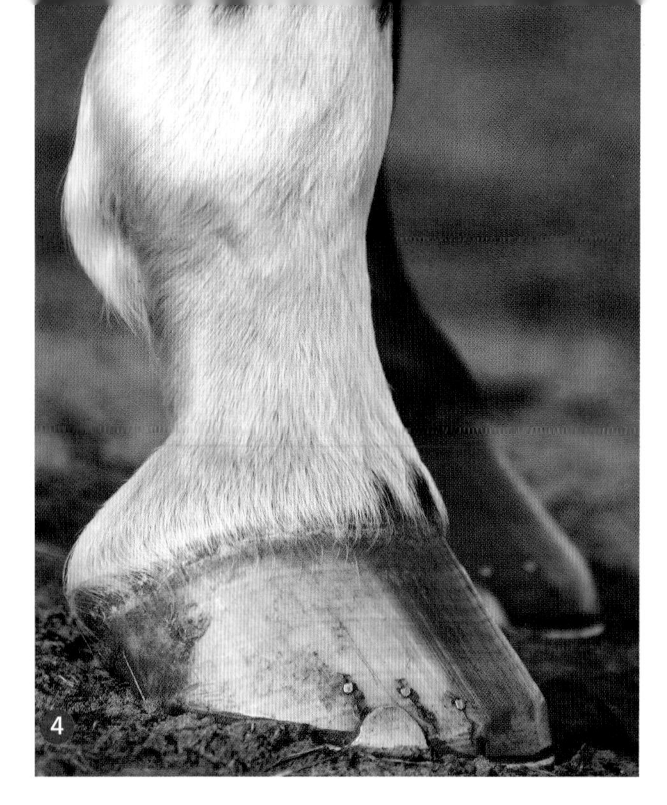

Multifunctioning hooves

The hooves of the horse (figure 4) have design features that produce just the right combination of strength, toughness, and cushioning. The foot bones give rigidity and strength, allowing a single leg to support very high loads. The horny hoof provides a hard, tough surface to protect the foot from knocks and scratches. The heel of the horse is highly elastic and is called a "frog," due to its soft cushioning characteristics. The frog helps provide grip and is also an aid to blood circulation. Shoe designers have spent many years developing new types of materials and layouts for shoes to improve the cushioning, grip and endurance. A well-designed hoof does not happen by chance.

Designer gears

Horses have a remarkable ability to walk and run with at least four different types of leg movement. This enables them to move with elegance and efficiency at different speeds. The most common "gears" of horses are referred to as walking, trotting, cantering, and galloping. As a horse speeds up or slows down it has the amazing ability to change smoothly from one gear to another and coordinate a different type of leg movement (figure 5).

Designed for endurance

The exceptional ability of the horse to run is illustrated by its ability to cover vast amounts of rough terrain in a day. Some horses compete in endurance events where they have to cover 100 miles in one day. In the Tevis Cup on the West Coast of the USA, the endurance challenge involves 100 miles of hilly and rough ground. Arabian horses are most often used in endurance races because they are the best at long-distance running.

4 The horse's hoof is carefully designed for multiple functions.

5 Example of trotting (left) and galloping (right).

6 Queen Elizabeth II was conveyed in a horse-drawn carriage during her Diamond Jubilee procession on June 5, 2012, in London.

7/8 Horses in the heat of battle. Figure 7 depicts Alexander the Great on his famous white stallion Boucephalus.

Horses for human use

God has tailor-made horses to be useful for people. They are perfectly designed for humans to sit on and they are valuable in many areas, including transportation, farming, policing, warfare, sport, and leisure (figures 6–8). When controlling crowds, a single police horse has the physical presence of about four policemen — and costs far less! Horses were once the most important form of transport and today they are still sometimes used to pull carriages (figure 6). During the industrial revolution horses were so well respected by engineers that the unit of power was called the "horsepower." The term "horsepower" is credited to James Watt in the late 1700s and is still used today.

History shows that horses have always been important to both armies and farmers. Historical pictures of wars often show horses as fearless creatures in the heat of battle. The bravery of horses is described by God in the Old Testament Book of Job: "He gallops into the clash of arms. He mocks at fear, and is not frightened; nor does he turn back from the sword" (Job 39:21–22). It is God who has designed the horse to be a brave, powerful, and valuable aid to humans.

Camels

Camels have some amazing features that are designed to help them cope with harsh desert environments, and they have been important domesticated animals for thousands of years.

1 Camels at a watering hole in Sede Boker Desert, Israel **2** Even the nose of a camel is designed to aid its survival in the desert.

Camels can cope with the extreme conditions of high temperatures, lack of water and vegetation, and a tough terrain. During the winter they can survive several months without drinking water, and even in the heat of summer they can go for several days without water. There are two types of camel: the Arabian camel with one hump and the Bactrian camel which has two humps.

Water system for the desert

Camels have a remarkable ability to find water through their fine sense of smell. When Earth is damp it produces an "earthy" scent caused by a chemical called *geosmin*. The camel can sense this damp earth from several miles away and, having found water, it can quickly drink and store up to 40 gallons in a single drinking session (figure 1)!

Camels even have special design features to conserve water. To reduce water loss they do not sweat until a higher temperature than most other mammals. Also, their kidneys produce concentrated urine and their urinary system is able to pass thick urine to minimize water loss. Camel dung is also very dry.

When an animal breathes out, the exhaled air contains water vapor that has come from the body. However, camels trap some of this water vapor with protrusions in the nose that are ideally shaped to condense water and return it to the body.

In dry countries, engineers sometimes use condenser machines to turn water vapor into liquid water for human use. The condensers consist of many intricate components in a complex assembly. Both man-made condensers and the camel's nose (figure 2) bear hallmarks of design.

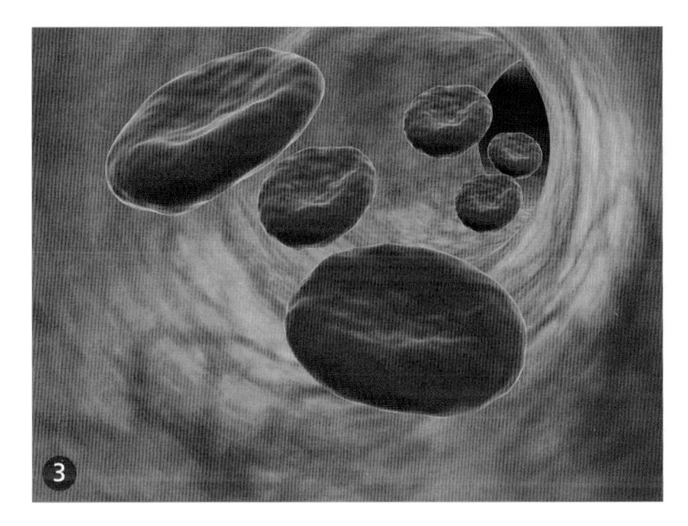

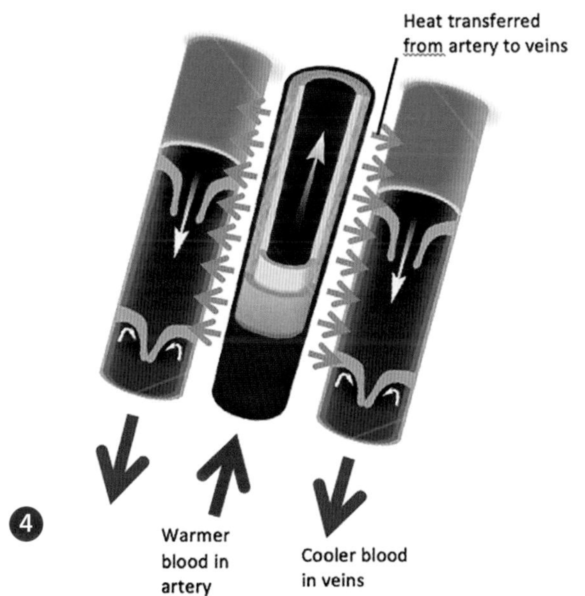

3 Camels have unique oval-shaped blood cells.
4 Counter-current cooling systems in blood vessels

Camels can survive extreme dehydration and cope with losing up to 40% of their bodyweight in water. This is due in part to oval red blood cells which can travel along the smallest of blood vessels (figure 3), even when blood thickens during times of dehydration. During rehydration the camel's red blood cells are also capable of expanding by up to 240% of their original volume without rupturing, whereas most animals' cells can expand only 150%. This makes it possible for the camel to drink large amounts of water to recover from dehydration.

Cooling system for the brain

Unlike most other mammals, the body temperature of the camel can vary by several degrees without the camel suffering any ill effects. For this to be possible, a camel must have a high-performance system for cooling the brain to keep it at a more constant temperature.

This brain-cooling system — called a *rete mirabile* (wonderful net) — works like this: many of the arteries and veins are located next to each other so that the hot arteries are cooled by the veins that are cooler because they are near the surface. Also, by flowing in the opposite direction, the cooling of the arteries is more efficient — this is called a counter-current system. And by having two veins next to the artery, the efficiency is higher still, as shown in figure 4.

Car engines are prevented from overheating by water that is continuously pumped through a circuit of channels that pass through the engine and a radiator where it is cooled down before it returns through the engine. This is far simpler than the camel's cooling system because the hot and cold pipes are kept separate and there is no counter-current system. One day engineers may attempt to copy the advanced design features of the camel.

Shaped for keeping cool

Animals generally have fat that is evenly spread over their body. However, the camel has most of its fat located in one or two humps on its back. One advantage of this is that the overhead sun does not have much area to heat. Another advantage is that the hump provides a layer that insulates from the fierce sun. As well as an insulating layer of fat on top of the body, camels have a thick fur coat. A fur coat is normally associated with insulating from the cold; however, the principle works exactly the same way when insulating from heat.

Non-sinking feet

The camel's foot is made up of two large toes with webbing in between to create a broad contact with the ground, as shown in figure 5. The toes are

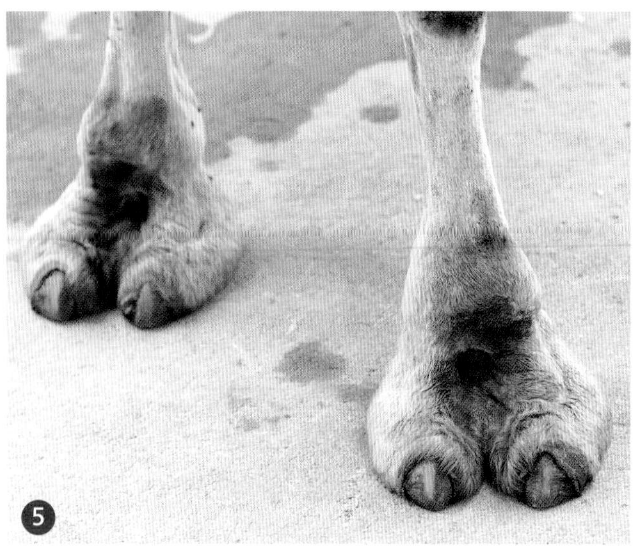

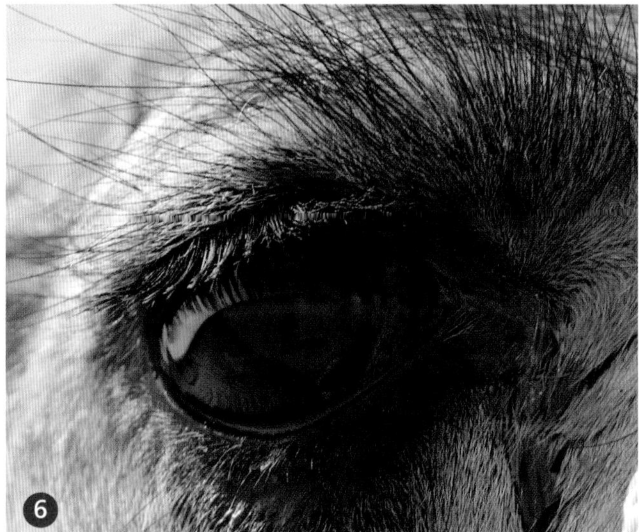

5 Camel feet are designed not to sink in soft sand. **6** A camel eye **7** A camel train or caravan

designed to spread apart under load to ensure that the foot does not sink into the soft sand. This is an important design feature because camels can weigh up to 2,200 lbs., and can carry loads of over 990 lbs. The foot also has a thick padded sole which protects it from being burned by the sand.

Engineers use continuous tracks (caterpillar tracks) to enable a vehicle to travel over soft sand. Caterpillar tracks could not have evolved from wheels but had to be specially invented by human ingenuity. In the same way, camels' feet were consciously designed.

Eyelids for a sandstorm

Camels have three eyelids. Two of the eyelids have eyelashes which help protect their eyes from sand (figure 6). The third eyelid is under the other two and goes from side to side like a windscreen wiper to remove sand from the eyes. This inner eyelid is clear, enabling the camel to see even when it is closed (also true of eagles – see p. 67). The camel can therefore keep its eyes open — even in a sandstorm!

Lips for tough vegetation

The thick rubbery lips of a camel enable it to eat hard prickly plants like cacti. Desert plants often have hard leaves with sharp spines in order to prevent animals from eating them; food is so scarce in the desert that plants would not survive unless they had such protective mechanisms. However, camels are one of the few animals that can eat such plants (figure 8).

Camels for human use

From very early times camels have been used for transportation. Tied together to form a "caravan," a dozen camels can be guided by just one or two people (figure 7). In addition, camel's hair is used to make carpets, rugs, and garments, and their skin is used for sandals, leggings, and water bottles. Camel milk is often an important source of liquid and nutrition for herders. Even the dung of camels is commonly used for fuel. Such wonderful provision is what would be expected from a Creator who designed everything on Earth for us.

Camels and the Bible

Camels were used widely in Bible times and were a sign of wealth. One of the earliest records of domesticated camels is found at the time of Abraham around 2000 B.C. (Genesis 12:16) and a seal depicting a Bactrian camel carrying a load has been found dating from around 1800 B.C. (figure 9). In the New Testament, Matthew records that John the Baptist wore a garment of camel's hair (Matthew 3:4). Camels may have been used by the wise men when they traveled to see Jesus. Jesus mentioned camels in His teaching that riches can be a stumbling block that keeps people from serving God: "It is easier for a camel to go through the eye of a needle, than for a rich man to enter into the kingdom of God" (Matthew 19:24).

8 A camel eating coarse desert vegetation

9 A seal dated around 1800 B.C. depicting a Bactrian camel carrying a load. One camel here has been highlighted to identify it. *(Courtesy British Museum)*

Elephants

Elephants are the largest living land animal, weighing up to 7 tons. Despite their great size they can perform delicate tasks with their unique trunks

1 A herd of African elephants drinking at a waterhole **2** An elephant using its trunk for a shower

There are two types of elephant today — African and Asian. African elephants are the larger, standing 10 to 13 feet high and weighing 8,800 to 15,400 lbs. African elephants also have larger ears. Elephants can live up to 70 years in the wild and are still widely used as working animals.

A unique trunk

Elephants are so tall and heavy that it would be very difficult and tiring for them to reach the ground with their mouth to eat food. Therefore they have a unique trunk that enables them to collect food and water from the ground with little effort (figure 1). The trunk is both skillful and powerful — it can crack open a peanut without breaking the seed, but it can also lift enormous loads of up to 770 lbs.

This unique trunk is a masterpiece of design consisting of around 40,000 muscles, together with a blood supply, nerves, and sensors. Even with the best current technology, engineers cannot fully replicate an elephant's trunk. Apart from collecting food and water, the elephant's trunk is so versatile that it can act as a snorkel when swimming, spray water over its body for cooling (figure 2), communicate by trumpeting, and smell and feel things through sensory hairs.

Ears for cooling and communication

Elephant ears have a large surface area with many blood vessels for cooling (figure 3). The cooling function is vital because elephants are huge animals and live in hot climates. When there is a cool breeze, they will spread their ears out into the wind to cool themselves down, but if there is no breeze they can flap their ears to create a cooling flow of air. Elephants also communicate with their ears — for example, vigorous flapping is a warning sign!

Long-distance hearing

Elephants can hear the call of another elephant up to 6 miles away because they are optimized for long-distance hearing. Their ears are tuned mostly to low frequency sound called "infrasound," which travels much farther than sounds humans can hear. They even use their feet to "hear" sounds by detecting seismic vibrations in the ground. When agitated, elephants stomp their feet to warn the whole herd.

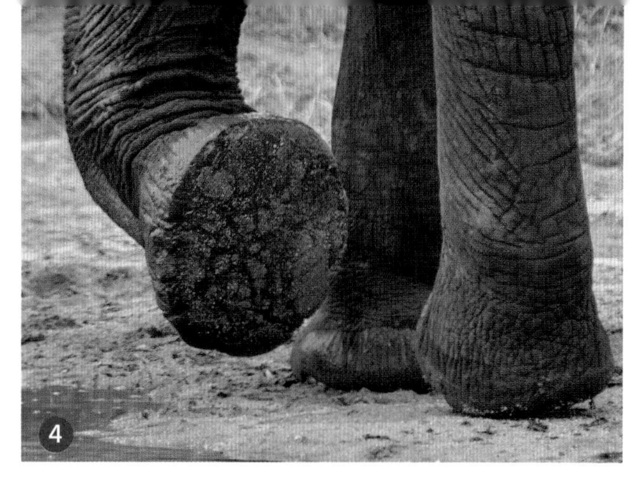

Designer feet

The tough fatty tissue that makes up elephants' feet acts as a strong, soft lining like a tire. This enables them to walk quietly in spite of their size and weight. Equally, they are able to walk in mud without getting stuck, because as the elephant lifts its foot out of the mud, it contracts, reducing suction. The sole of the foot has ridges which help the elephant grip muddy terrain in the same way that tractor tires grip the mud (figure 4).

Big stomach for a big appetite

An adult elephant has a huge stomach that can hold over 220 lbs. of food in a single meal! Elephants are able to live in harsh environments because they can digest low-quality vegetation that is high in fiber and low in protein.

Elephants for human use

Elephants have been excellent working animals for humans throughout history, and they were sometimes used as a "tank" in ancient warfare. They can be trained to follow over 30 commands, including lifting their leg or trunk to enable people to climb up onto their back (figure 5). They are useful for hauling heavy loads in remote areas and for transporting people. An elephant ride is not only smooth, but there is also a great view from such a highly elevated position (figure 6)!

3 The large ears of this South African elephant aid its cooling.

4 An elephant's foot showing the tread pattern

5 Mounting an elephant using the trunk

6 An elephant ride is a high-level experience.

Giraffes

The giraffe is the tallest land animal in the world with an adult height that can be over 16 feet. There are some very specialized design features in a giraffe's body that enable it to be so tall.

1 A group of reticulated giraffes in Buffalo Springs National Park, Kenya
2 The giraffe neck can bend right around to reach its body for grooming.

Giraffes live primarily in African savannas, which consist of rolling grasslands with scattered bushes and trees. Their extreme height allows them to eat leaves and food located much higher than other animals can reach (figure 1). They also use their height to watch out for predators.

Unique neck for reaching high and low

The giraffe has a unique and powerful neck that can reach 8.2 feet in length. Like most mammals, giraffes have around 7 vertebrae in the neck, but these bones are unique in both size and design in the giraffe. The vertebrae join to each other with a ball and socket joint so that the neck has a combination of smooth motion and a large range of movement. The neck enables the head to be lowered to the ground and also to bend sideways (figure 2). For its head to reach the ground, a giraffe either does the splits (figure 3) or bends the knees of its front legs.

High performance cardiovascular system

The giraffe's heart has to have enough power and pressure to handle more than 16 feet of gravitational resistance. To overcome this challenge, giraffes have a large heart of up to 2 feet in length and a blood pressure about twice that of humans. Giraffes also have narrow capillaries and small blood cells that allow for faster absorption of oxygen.

When the head is lowered, this creates a tendency for blood to rush to the head. To prevent this, special shunts in the arteries supplying the head restrict blood flow to the brain, diverting it into a web of small blood vessels (the *rete mirabile* or "marvellous net"). Also, valves in the jugular veins prevent returning blood from flowing backward into the head while the head is lowered.

Tough and tight skin

The giraffe's skin is thick to allow it to run through thorn bushes. The skin is also very tight, especially around the legs, and this aids blood circulation in the same way that compression tights improve circulation in humans. The giraffe has been studied by engineers trying to produce better astronaut suits for space where the lack of pressure creates a need for tight-fitting spacesuits.

A tough tongue

The giraffe's tongue can be up to 19 inches long and it is very prehensile, which means it can wrap around leaves to tear them from the branch (figures 4 and 5). The tongue is a dark purple color, which helps protect it from sunburn. A full-grown giraffe can consume almost 100 lbs. of leaves and twigs a day. The lips, tongue, and inside of a giraffe's mouth are covered in small, tough growths called *papillae*, enabling them to eat thorny plants. Thick saliva also helps protect the mouth.

Growing up fast

A giraffe can be 6 feet tall at birth and can double in height by its first birthday. It can stand within a few minutes of birth and walk within an hour. After just 24 hours the young giraffe can jump and run (figure 6). Such an amazing creature bears testimony to a Creator of infinite wisdom.

3 A giraffe often spreads its front legs out to reach the ground.

4 A giraffe's tongue can wrap around leaves to gain a good grip for pulling.

5 A giraffe uses its tongue to reach even higher leaves.

6 A calf jumping a few days after birth

Kangaroos

Kangaroos are the only large land animal that moves mainly by hopping. They live in Australia and some of the surrounding islands, and are able to survive intense heat and long exposure to the sun.

1 An Australian gray kangaroo

Elastic joints for hopping

Kangaroos live in dry, open landscapes (figure 1) and need to be able to travel long distances in order to find food and water. Hopping is an efficient means of travel because the joints in the kangaroo's legs act like springs. When the kangaroo lands on the ground during hopping (figure 2), the tendons in the hind legs stretch like an elastic band. Just after landing, the tendons contract to release energy to help power the next hop (figure 3). The movement of the body as it hops helps the kangaroo breathe, making it even more efficient. Engineers have created hopping robots using some of the best technology available in engineering — and yet these robots are far inferior to a real kangaroo.

Strong tail for balance and standing

Kangaroos have a very large tail which is used for balancing during hopping by forming a counterweight to the body as shown in figures 2 and 3. A kangaroo can stand upright because the large rear feet and large tail form a stable tripod structure (figure 4). Standing up enables the kangaroo to get a better view. It also enables male kangaroos to box each other! The tail is so strong that the kangaroo can even support itself momentarily on its tail in order to kick an opponent (figure 5).

A cozy pouch for the young

A kangaroo is a *marsupial* mammal, which means that the offspring live in a pouch for some time after birth. When the "joey" is born, it climbs to the teat in the mother's pouch where it stays for around nine months before starting to leave for short periods until it is weaned. When one joey is out of the pouch but not weaned, the mother can have another joey still attached to a teat, and an embryo not yet born. She is able to delay the development of that embryo until the older joey leaves the pouch; or if there is a drought, until there is more food available. At the right time, her body sends hormones to restart the growth in the womb.

Evaporative cooling

Kangaroos have a remarkable ability to survive in temperatures of up to 122° Fahrenheit. One way they cool down is by licking the skin on their legs to create evaporative cooling. When water evaporates

it uses up latent heat which cools the skin. Kangaroos have a network of blood vessels in their legs that make this efficient and, when the temperature is high, more blood flows to that area. They also lose heat by panting and sweating.

Self-repairing skin

Kangaroos have a special enzyme that repairs DNA in the skin that has been damaged by being exposed to bright sunlight for long periods. This means that the risk of skin cancer is reduced. Research is being carried out to see whether these enzymes could help reduce skin cancer in humans.The kangaroo shows that God can create an animal for the most extreme environment. It also reveals God's incredible variety of design solutions to the particular needs of animals.

2 During landing, the large rear legs store elastic energy.

3 Elastic energy is released during takeoff to help power the next hop.

4 With long feet and a long tail, kangaroos are perfectly balanced for standing.

5 Kangaroos have a unique ability to (momentarily) stand on their tail.

6 A female can have more than one joey in different stages of development.

Sheep

Sheep are one of the most common domesticated animals with a world population of around one billion. They can be kept almost anywhere where there is grass, including mountain locations.

1 A sheep on the Great Ridge in the Peak District, England *(Photo: Brian Edwards)*

2 Sheep are normally shorn once a year in the spring.

There are several hundred breeds of sheep that have been developed for different uses. Some are bred mainly for wool (figure 1), others mainly for meat or milk or simply for managing grass.

High-performance wool

Sheep can be bred to produce wool fiber from very fine to coarse. The fine wools tend to be used for clothing, while coarse wools are used for carpets. Despite decades of research on man-made fibers like polyester, wool is often considered a superior material for clothes and carpets due to its better insulation and better moisture absorption and the fact that it is more durable, more fire resistant, and more wrinkle resistant.

Sheep's wool grows fast enough to give a sizeable annual crop of wool (figure 2). Cutting the wool once a year is very convenient in countries where there is a cold winter because the sheep can have a thick coat in the winter months and then be shorn in the spring. This means they can live in cold and remote mountain areas (figure 3).

Staying fresh

Wool is less prone to giving off smelly odors than man-made materials like polyester and nylon. This is because a natural wool fiber is highly textured with scales whereas a man-made fiber like polyester has a much smoother surface. The intricate surface of wool creates stronger bonds with odor molecules and stops them escaping from the surface and creating an unpleasant smell.

Eco-friendly grass management

Some breeds of sheep are ideal for keeping grass short and controlling weed growth between trees in orchards and tree plantations (figure 4). These sheep also help to reduce the spread of fungal diseases by consuming fallen leaves. Whereas many breeds of sheep strip bark and foliage from trees, some breeds, like the Shropshire, have a well established reputation for being "tree safe."

Sheep products

It is well known that lamb is a popular red meat around the world. What is not so well known is that sheep's milk is ideal for making cheese (figure 5) that is generally more nutritious than cheese made from cow's milk, because sheep's milk normally contains higher levels of calcium and vitamins A, B, and E than milk from cows. Famous types of sheep's milk cheese include French Roquefort, Greek feta, Spanish manchego, and Italian pecorino romano.

As well as providing us with wool, meat, and cheese, sheep are used in the manufacture of many other products (figure 6). Sheep gut can be used to make tough fibers for applications like tennis racket strings; tallow from sheep's fat is used to make candles and soap; and lanolin, a moisturizer extracted from wool, is used to help make cosmetics and skincare products.

Sheep and humans

In Psalm 50:10 God told his people that "Every beast of the forest is mine, and the cattle on a thousand hills," which reminds us that not only are the wild animals created by God, but the domesticated animals are purposefully designed for human use.

3 Some breeds of sheep are hardy enough to live in rough mountainous areas. *(Photo: Brian Edwards)*

4 Sheep can be a better solution than a mower for managing grass. **5** Cheese made from sheep's milk

6 A few of the products that use materials from sheep.

Dogs

Dogs are very intelligent creatures and have one of the best senses of smell of any land animal. They carry out many important jobs for humans and also make great pets.

Although dogs are found in the wild, most dogs alive today are either working dogs or pets. Over hundreds of years they have been bred into a variety of sizes, abilities, and temperaments. Working dogs include sheepdogs, guard dogs (figure 1), sniffer dogs (figure 2), sledge dogs, and guide dogs.

Designed for smelling

The sense of smell is called "olfaction" and happens through sensory cells in the nose. When odorous molecules (pleasant or unpleasant smelling) make contact with scent receptors in the nose, a signal is sent to the brain and a smell is sensed. The area in a dog's nose with scent receptors is more than ten times larger than the equivalent area in humans. A dog also has up to a hundred times more receptors per square centimeter. Its nose is wet, which makes smelling more efficient by trapping airborne particles.

Remarkably, when dogs breathe in, a fold of tissue just inside their nostril splits the airflow into two different pathways — one for smelling and one for breathing. Dogs can wiggle their nostrils to determine which nostril a scent arrived in and this helps locate the source of smells. The part of the brain that processes smells is 40 times larger in a dog than in humans, showing that they have the computer processor, not just the sensors (hardware), to detect and analyze smell.

1 Dogs were domesticated early in the history of the human race and with good reason they have been called "man's best friend." *(Photo: Brian Edwards)*

Sniffer dogs

Dogs are widely used by police and other security forces for sniffing out specific items like explosives, firearms, drugs, and illegal foods. They are also used to find valuable foods in the wild like truffles, which are the fruiting body of an underground fungus and an expensive food delicacy.

Bloodhounds have around 300 million scent receptors which is more than any other breed. They not only can follow a scent on the ground (figure 3), but they can also follow it in the air. Their whole head is specifically designed for tracking: they have a long head, a nose with large nostrils, long ears that sweep the scent up from the ground, and a cape of loose skin around the head and neck to trap and retain the scent.

Designed to run fast

Unlike humans, dogs have no collarbone, which gives more flexibility to the shoulders and hence allows them to have a greater stride length. Dogs also have a spine that is flexible, allowing them to bring their back legs forward when running (figures 4 and 5). The powerful muscles in their back legs enable fast running and high jumping. Some dogs can jump three times their own height.

Designer insulation

Every part of the body of Siberian huskies is well insulated (figure 6). They have two coats of fur with an undercoat of warm dense fur and an overcoat of longer fur that is water-resistant. They have hairs inside their ears and almond-shaped eyes to allow them to squint in order to shut out the cold winds. Their long and bushy tail can curl around the nose to warm the air around the face while the dog sleeps.

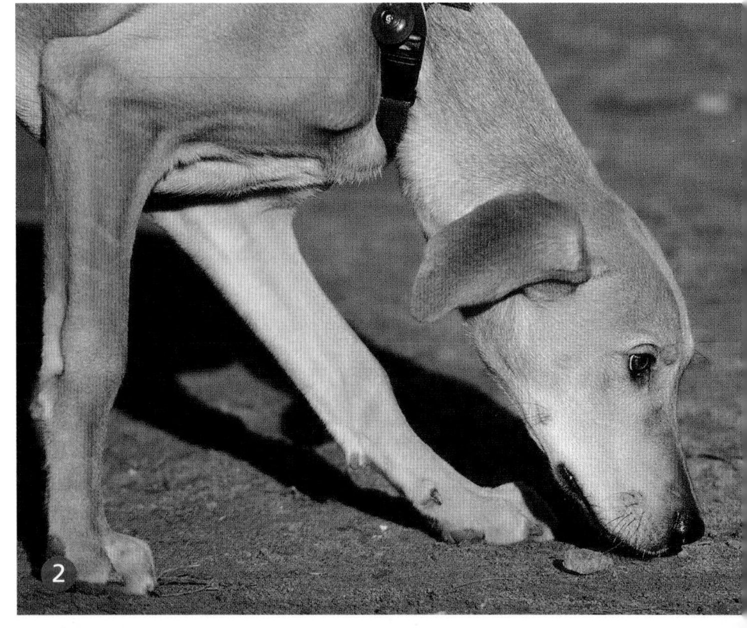

2 A dog's sense of smell is more than a thousand times more sensitive than that of humans.

3 Bloodhounds are among the best scent-trailing dogs.

4 Greyhounds are the fastest dog breed and can run up to 40 mph.

5 Malamutes are powerful dogs that can be used to pull heavy sledges.

6 Siberian huskies can withstand temperatures as cold as -58° Fahrenheit.

They also have thick, furry footpads that keep their feet insulated.

Designer strength

In areas with a lot of snow, dogs are used to pull people and other loads on sledges. Huskies and malamutes are powerful dogs that are commonly used for sledge-pulling (figure 5). A single adult male Alaskan malamute can pull over 2,200 lbs. of weight across snow.

Medical assistance dogs

The calmness, intelligence, and alertness of trained dogs make them very good at assisting people in need (figure 7). Dogs can be trained to be guide dogs for the visually impaired, and for the hearing impaired they are trained to react to certain sounds such as alarms, telephones, or the doorbell. Dogs can also assist people with mobility problems. These dogs are taught to do practical jobs like picking up objects that have been dropped, or getting help in an emergency.

An instinct for herding

Dogs can be used for herding sheep (figure 8), cattle, goats, reindeer, and even geese. Some breeds, like border collies, have a natural herding instinct and are easy to train. For herding, it is estimated that a sheepdog can do the work of three humans. The handler uses whistles, words, or hand signals to instruct the dog. These sheepdogs can perform complex tasks such as splitting a group of sheep in two and going back for a lost animal.

7 Pets As Therapy (PAT) is a U.K. charity providing therapeutic visits of dogs to hospitals, hospices, nursing, and care homes.

8 Dogs are particularly suited to herding sheep. 9 There are estimated to be over two hundred million pet dogs in the world.

Police dogs

As well as searching for dangerous items like explosives, police dogs are trained to track and chase crime suspects. They can disarm violent suspects and help control hostile crowds. They can also be trained to search for missing people. After the World Trade Centre attacks in September 2001, dogs were used to locate trapped people.

A brain for learning

Some of the reasons why dogs are so useful to humans are their intelligence and ability to be trained. Dogs have been known to be able to follow over 100 different commands. Some dogs learn to carry out a command after only five repetitions. They can respond to spoken cues, whistles, or hand gestures. Some breeds are easy to train because training can be included in their play, or because they are food-focused and so are motivated to obey. Many dog owners train their dogs as a hobby because they and their dog enjoy the stimulation. A well-trained and obedient dog is an asset to any owner.

Designer pets

Dogs make wonderful pets due to their intelligence and willingness to please. They can be chosen on the basis of size or temperament or the amount of exercise or grooming needed. Whatever the age and family situation, there is usually an appropriate breed of dog for a particular home (figure 9).

The remarkable intelligence and trainability of dogs show that they were purposefully designed to be a special help and friend to man. Even though all dogs (e.g., wolves, coyotes, dingoes, and domestic dogs) appear to be descended from an ancestral type of dog that looks like a wolf, it is nevertheless true that God designed the great genetic diversity in the first dog kind that enabled human breeders to produce a great range of dog breeds. So it is still appropriate to give thanks to God for all the different breeds of dog that are a help and comfort to mankind today.

03 Sea Creatures — Coral Reef Fish

Coral reefs are among the most beautiful places on Earth containing a spectacular array of fish, many of which have striking shapes, patterns, and colors.

1 A coral colony together with tropical fish and sea turtles in the Red Sea, Egypt

Gardens of the sea

Coral reefs are places of immense beauty — they are the gardens of the sea. Figure 1 shows a coral reef in the Red Sea with crystal clear water and an abundance of colorful plants and creatures. Coral fish often live in large shoals, like the shoal of blue-lined snappers shown in figure 2. Coral reefs are made up of small marine invertebrates called corals. The corals are stationary and cluster with other corals to form colonies. The colonies then secrete calcium carbonate which causes them to bind together to form the reef. Each of these individual coral animals is covered with a hard exoskeleton that makes coral reefs appear rock-like. The crevices and caves provide shelter for coral fish.

Coral reefs grow mostly in warm, shallow, and agitated waters. They are often found around islands and coasts near to the equator. The largest coral reef is the Great Barrier Reef off the coast of Queensland, Australia, which is 1,553 miles long (figure 3). The second largest reef is the Red Sea Coral Reef near Egypt and is 1,180 miles long. Coral reefs occupy less than 1% of the surface area of the world's oceans, yet they contain 25% of all marine species. They are home to fish, sponges, corals, jellyfish, worms, shrimps, spiny lobsters, crabs, starfish, sea urchins, sea cucumbers, sea squirts, sea turtles, and sea snakes. There are about 6,000 to 8,000 species of fish around the reefs with a great range of sizes, shapes, and colors.

Designed for maneuverability

Open water fish like salmon have a streamlined shape for speed and efficiency. In contrast, coral fish often have a flat shape, like the threadfin Butterflyfish shown in figure 4, which enables them to exert large sideways forces and hence turn quickly. Many coral fish hide in crevices in the reef (figure 5). The same hideouts may be inhabited by different species at different times of day. Night-time predators such as cardinalfish and squirrelfish hide during the day, while daytime fish such as surgeonfish, triggerfish, wrasses, and parrotfish hide during the night.

Special relationships

When two creatures depend on each other for existence this is called symbiosis. The bluestreak cleaner wrasse is an example of a fish that has a symbiotic relationship with larger fish by eating parasites and dead tissue on their skin (figure 6). This gives health benefits to the larger fish as well as providing food for the cleaner wrasse.

Another special relationship exists between carrier crabs and sea urchins. The carrier crabs put the sea urchins on top of them to look scary. In return, the sea urchins benefit from being transported to new feeding grounds.

Another symbiotic relationship exists between sea anemones and clownfish (figure 7). A sea anemone

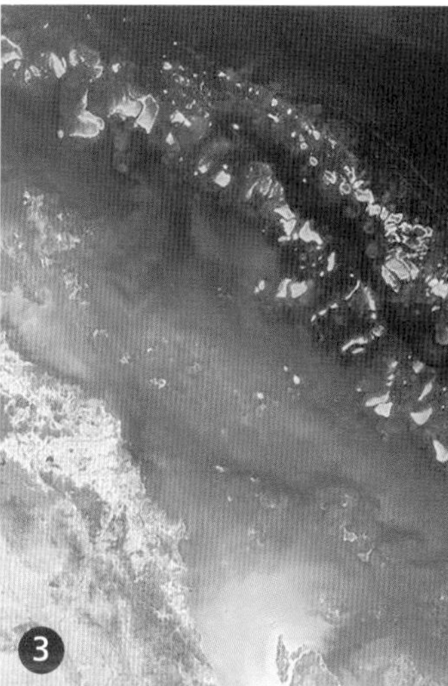

2 A shoal of blue-lined snappers in the Maldives

3 The Great Barrier Reef near Australia

4 Threadfin Butterflyfish in the Red Sea

5 Tiger fish hiding in a crevice

6 Giant moray eel being cleaned by a bluestreak cleaner wrasse

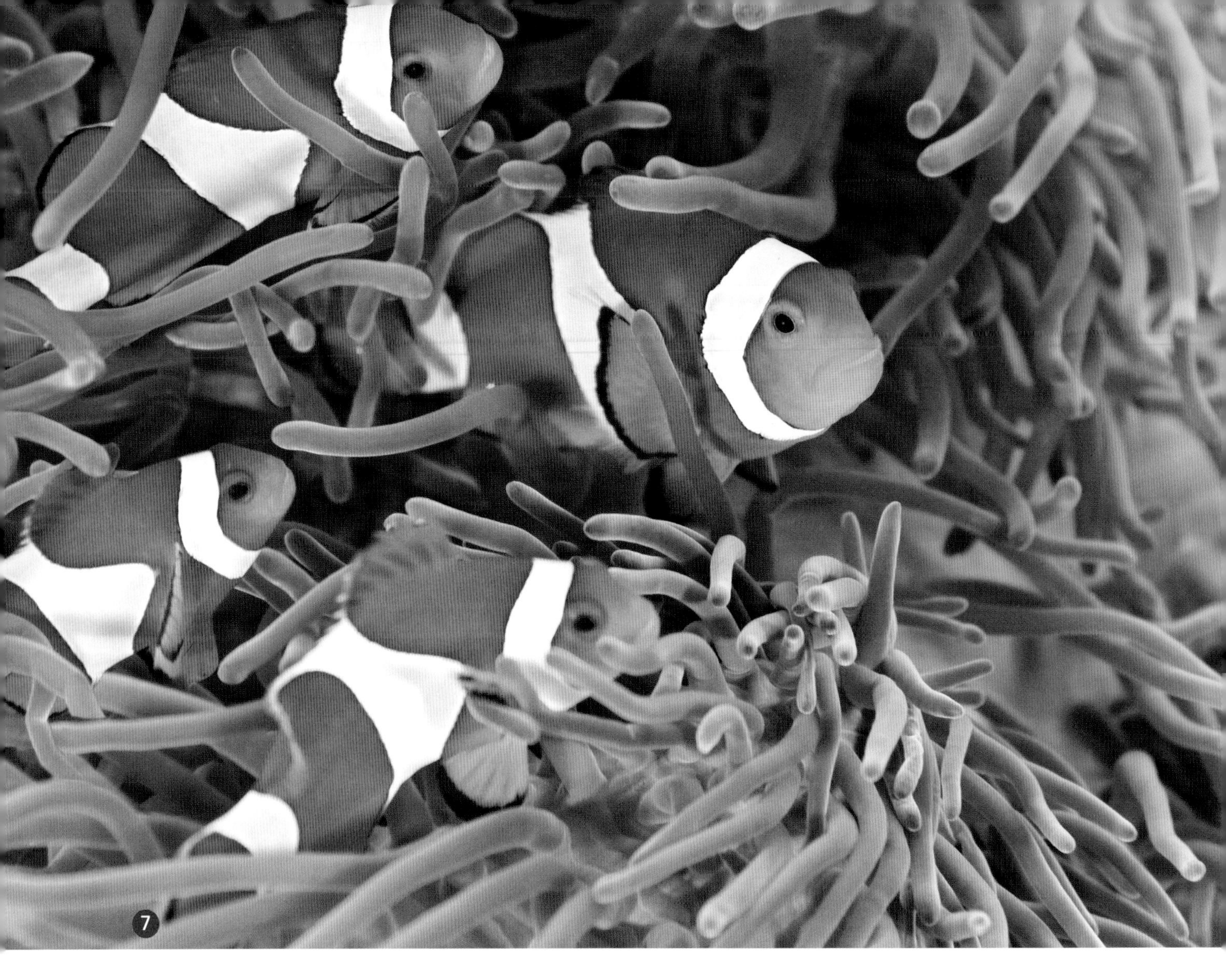

7 The sea anemone provides protection for the clownfish.

makes an ideal home for a clownfish because the poisonous tentacles of the anemone provide protection from predators, and leftover food from the anemone is eaten by the clownfish. In return, a clownfish can help an anemone catch its prey by luring other fish toward the anemone. Clownfish even eat dead tentacles which helps to keep the anemone clean.

Beautiful coloring

Coral reef fish exhibit a dazzling variety of colors and patterns (figures 8–13). They have special cells called chromatophores that can produce bright coloring; they can also switch colors off and on and even produce bioluminescence. Bioluminescence is where organisms produce their own light by way of a chemical reaction or by harboring bacteria that produce bioluminescence.

It is interesting to note that when fish have a protective environment like a coral reef, they often have bright coloring. In contrast, fish in open spaces or open waters usually have dull colors because of the need for camouflage from predators. This indicates that in the original creation, when there were no predator-prey relationships, bright and beautiful colors were the norm. Beautiful features like bioluminescence are extremely complex and yet such features are often not needed for survival. The coloring of coral reef fish is another example of beauty for the sake of beauty which can only be explained in terms of intelligent design.

The beautiful kaleidoscope of colors among fish in a coral reef.

8 Redlion fish **9** Mandarin fish **10** Copperband butterflyfish **11** Redbreasted wrasse

12 Pajama cardinalfish **13** Sweetlip fish

Whales

Whales are beautiful because of their great size, graceful swimming, and even their singing. They are the largest creatures on Earth.

1 Inside the mouth of a humpback showing the baleen filters **2** A gray whale showing the baleen filters

There are two main categories of whale— toothed and baleen. Toothed whales include dolphins, killer whales, belugas, and sperm whales. Baleen whales include blue whales, grey whales, and humpback whales. Baleen whales do not have teeth but instead have baleen plates in the upper jaw which act as filters for trapping food. Figure 1 shows the baleen plates of a humpback whale and figure 2 shows the baleen plates of a grey whale. This section deals with baleen whales, which are generally the larger whales.

Big appetites

The blue whale is the largest whale with a length of up to 98 feet and weight of up to 396,000 lbs. Figure 3 shows a schematic of a blue whale next to a human to illustrate the immense size of the whale. Blue whales eat about 8,800 lbs. of krill each day in the feeding season. Newborns are around 20 feet long and drink 50 gallons of milk a day! The diet of other baleen whales includes krill, small fish, and plankton. Baleen whales can take a mouthful of prey and force the water back out, while trapping the prey inside the baleen plates. These plates are made of keratin (the same as fingernails) but are unlike any structure in any other animal.

Big is helpful

Their size gives whales an advantage in cold water as they lose less heat due to their low ratio of surface area to volume. This is important because whales are warm-blooded. A large amount of blubber also acts as insulation and a food store. Size also helps whales to swim fast because larger sea creatures swim faster due to larger fins.

The immense size of whales is possible because of the way water supports their body in all directions. Even though they are so heavy, their skeletons are relatively light.

Jumpers and divers

Some whales, like humpbacks, can jump out of the water, which is called breaching (figure 4). Scientists are puzzled why whales do this because it takes a lot

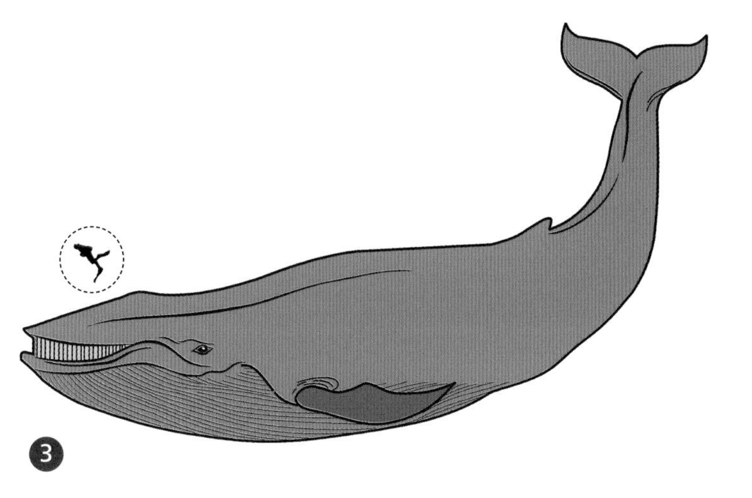

3 The blue whale is the largest of all sea creatures and a 6-foot-tall man would be a mere 1/16 its length!

4 A humpback breaching

5 Raising the tail to provide impetus to the dive is called fluking.

6 A humpback feeding at the surface

of energy to make such jumps. It has been speculated that whales are communicating something to their peers. However, an alternative explanation is that God wanted whales to make spectacular displays for his pleasure and the pleasure of human observers.

Sperm whales can dive to an incredible depth of up to almost 10,000 feet, which is far more than the human free-diving limit of around 656 feet. When diving, the whale's lungs collapse under pressure, which keeps them from rupturing. At the start of a dive, whales sometimes lift their tail out of the water to aid diving. This is called fluking and is shown in figure 5. One of the reasons why whales dive straight down is to approach shoals of small fish from below. Figure 6 shows the mouth of a humpback rising out of the water.

Singing whales

Male humpbacks have a remarkable ability to sing songs with musical structure such as rhythm, key signature, and melodic phrases. The songs can last 20 minutes and consist of several musical themes. To

compose a melodic song that lasts 20 minutes involves selecting hundreds of notes with precise musical characteristics. However, whales are not educated in musical structure and composition! Whale songs are an example of added beauty that can only be explained in terms of intelligent design. The Bible teaches that God has made innumerable creatures in the sea both "small and great" (Psalm 104:25). Whales certainly come into the "great" category.

Dolphins

Dolphins are designed to swim with speed and agility. They are also very intelligent creatures which can be trained by humans.

1 A bottlenose dolphin mother and her calf **2** A pod of spinner dolphins

Dolphins live mainly in the oceans but some spend time in freshwater rivers and lakes. They typically eat small fish and squid. The calves suckle milk from their mother for up to 18 months and swim close to her during this period (figure 1).

Dolphins live together in groups called "pods" where they play and hunt together. One fishing technique used by dolphins is to blow bubbles to trap fish in a tight group. Figure 2 shows a group of spinner dolphins which are so called because of the way they spin when they jump out of water.

A body designed for speed

Bottlenose dolphins are capable of bursts of speed of up to 30 mph and can dive down to depths of up to almost 1,000 feet below the surface of the ocean. The body of a dolphin has a teardrop shape which gives low drag. The outer layer of a dolphin's skin is shed and replaced every two hours to keep the skin smooth.

When engineers first studied dolphins in the 1930s they calculated that they should not be able to swim at the speeds they do considering the quantity of muscle in the body and the resistance of water. Engineers called this phenomenon "Gray's paradox" after the engineer (Sir James Gray) who discovered it. In recent years, engineers have been studying dolphins closely for inspiration to design better ships.[1]

Designed for acrobatics

Whereas fish have vertical tail fins, dolphins have horizontal tail fins. Such a tail fin enables them

1 Frank E. Fish, "The Myth and Reality of Gray's Paradox: Implication of Dolphin Drag Reduction for Technology." *Bioinspiration and Biomimetics*. 1 (2006) R17–R25

to be adept at jumping out of the water, which is called "porpoising" (figure 3). In contrast to fast fish like tuna, which are not very agile, dolphins have two powerful flippers for making sharp turns in the water.

Dusky dolphins (figure 4) are one of the most acrobatic dolphins. They can perform a number of aerial displays including leaps, backslaps, tailslaps, spins, and head-over-tail leaps.

Long range communication

Dolphins make a variety of sounds, including clicks, whistles, buzzes, and squeaks. These are used for communication, navigation, and locating predators or prey. Communication ranges of over 3 miles have been measured for bottlenose dolphins.[2] Dolphins produce high frequency sounds that reflect off objects and they then pick up the echoes. This echolocation enables them to recognize the identity and distance of other creatures.

Dolphins and human use

Dolphins can be trained to interact with humans in aquariums (figure 5). The U.S. Navy has trained dolphins to perform complex tasks like detecting underwater mines (figure 6).

2 Jensen et al.: "Bottlenose Dolphin Short-range Communication," *J. Acoust. Soc. Am.*, Vol. 131, No. 1, January 2012.

3 A dolphin family porpoising off the coast of Hawaii **4** A dusky dolphin porpoising off the coast of Kaikoura, New Zealand

5 Dolphins can be trained to play with children.

6 Dolphins carried out mine clearance work for the U.S. Navy in the Persian Gulf during the Iraq War.

Salmon

Salmon have an amazing instinct and determination to navigate vast distances in order to return to the exact stream where they were born.

1 Atlantic salmon **2** A cross section of salmon showing the muscles for swimming

Designed for stamina

Salmon have the stamina to travel long distances at sea and then to travel back to the stream where they were born for mating and spawning. Chinook salmon spend up to eight years in the Pacific Ocean traveling thousands of miles before returning to the stream where they hatched.

Salmon are very efficient swimmers with highly streamlined bodies (figure 1) and a powerful tail fin for propulsion. Another key reason for the speed of fish like salmon is that almost every muscle in the body is dedicated to swimming. Figure 2 shows how the central section of a salmon — which holds the organs like the stomach, intestines, and heart — is completely surrounded by muscle. The pink color of salmon muscle is partly due to the mixing of slow (white) and fast (red) twitch muscles and partly due to pigments (such as carotenoid) obtained through sea food. Salmon is an excellent source of food because it is up to 70% muscle. The meat is nutritious, low in saturated fat, and delicious!

Precision navigation and brave jumpers

Like sea turtles, salmon use magnetic sensors to navigate across the oceans. Once they have navigated to the right section of coastline, they use their sense of smell to direct them through the river system to get back to their place of birth.

Salmon have an incredible ability to sense chemicals to one part per million. This enables them to navigate a river system that may involve hundreds of miles and many junctions, and they often battle upstream against strong currents and waterfalls. Salmon usually stop eating once they get into the river system so they carry out their remarkable journey with limited energy reserves. They are expert jumpers (figure 3) with some species being able to jump over 13 feet in height. In Alaska

and Canada, salmon even have the extra challenge of grizzly bears to contend with (figure 4).

The next generation

Mature salmon return from the ocean to the spawning beds from where they hatched (figure 5). The female digs a hole in the gravel with her tail and lays between 1,000 and 17,000 eggs (figure 6), which are then fertilized by the male fish. The mass of eggs is called caviar and is an expensive delicacy. The young salmon usually hatch in early spring.

Fully programmed at birth

Baby salmon (called alevins) live under rocks and absorb nutrients from their yolk sac, which stays attached to them. Alevins grow into fry (figure 7) and live in the stream where they were born. After about one year the small fry follow the river to the ocean where they live in "schools." Finally, adult salmon go farther out to sea to live their lives, mostly alone, before returning to spawn.

Young salmon survive without any help from their parents. They have to swim downstream, navigating all the tributaries to find an ocean they do not know exists. When they reach the ocean they then mature over several years, instinctively knowing when to return to mate and spawn. Salmon are yet another example of incredible design that defies evolution. In particular, the young could never have gradually evolved the ability to perform such a complex migration because they cannot survive unless the whole migration cycle is fully programmed in their brain.

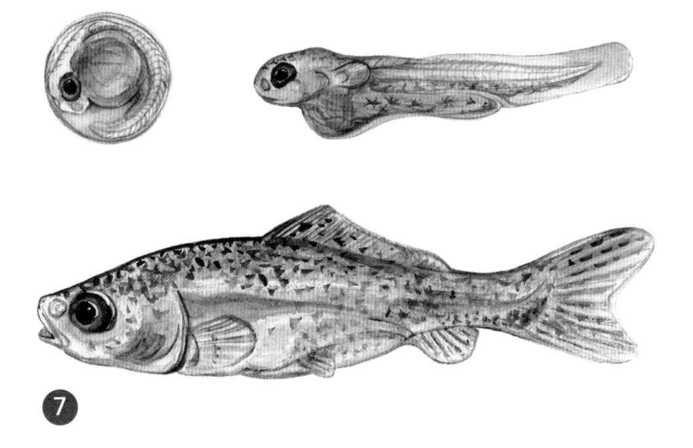

3 Salmon jumping at waterfalls **4** Salmon jumping past grizzly bears

5 Chinook salmon (Pacific salmon) spawning in a stream **6** Salmon caviar **7** Young salmon, known as alevins

Sea Turtles

Sea turtles are reptiles that live in the sea and lay eggs on land. Newly hatched sea turtles have a remarkable ability to find the sea and survive without the help of their mothers.

1 Sea turtles are usually slow swimmers but can have bursts of speed of up to 20 mph.
2 Leatherback turtles are the largest species of sea turtle.

Sea turtles are reptiles that are found mainly in warm sea waters. Their diet includes plants, algae, crabs, shellfish, and sometimes jellyfish. Sea turtles use flippers for swimming (figure 1), whereas fresh water pond turtles use webbed feet. Leatherbacks are the largest sea turtles (figure 2) and can reach up to 7.2 feet in length and 1,500 lbs. in weight.

Designed for the oceans

Turtles have two nostrils that are near the top of the head (figure 3) to make it easier to breathe during swimming. They have a slow metabolism which enables them to hold their breath underwater for long periods. Green sea turtles (figure 4) can hold their breath for up to five hours! Sea turtles have many rod cells in their retinas and this gives them good vision in the dark. This is important not just for night swimming but also for finding a nesting place on the beach at night. Sea turtles have beautiful shells that give physical protection from predators as well as camouflage. The shells are light and streamlined for efficient swimming.

Magnetic navigation

Sea turtles migrate vast distances across the oceans. Leatherback turtles have been observed to migrate over 10,000 miles, crossing the entire Pacific Ocean from Indonesia to the USA. Scientists have discovered that turtles navigate using magnetic sensors to determine latitude and longitude. Research has shown that newly hatched turtles use these magnetic signals to head toward the open ocean immediately after they reach the water.

Some sea turtles use a form of hibernation instead of migrating to warmer areas. They are able to draw oxygen from the water and their heart rate slows down.

Hatching all alone

After laying her eggs (figure 5) and covering the nest (figure 6), the mother returns to the sea. This means that when the eggs hatch, the young turtles have to survive alone and they may never meet their parents. They dig themselves out of the sand and make their own way to the sea (figure 7) by heading toward the brightest place. Despite the power of the waves and the difficulty of breathing for such tiny creatures, they instinctively head for the sea and the open ocean. These tiny creatures then follow their instincts for swimming, navigating, hunting, mating, migrating and, in the case of females, returning to the same beach where they were born to lay their eggs.

Like baby salmon, turtles are born fully programmed so they know how to live their entire lives, including complex migrations and reproduction. As with salmon, such a complex lifecycle cannot evolve step by step but must be fully designed from the beginning.

3 Turtles have nostrils near the top of their head.

4 A green sea turtle showing the beautiful shell pattern

5 Turtles lay up to 200 eggs.

6 A green sea turtle covering her nest on the beach

7 A newly hatched turtle about to enter the sea

Sea Snails

Sea snails are one of the most abundant and diverse sea creatures in the oceans. Their hard shells can be extremely beautiful with intricate shapes, patterns, and colors.

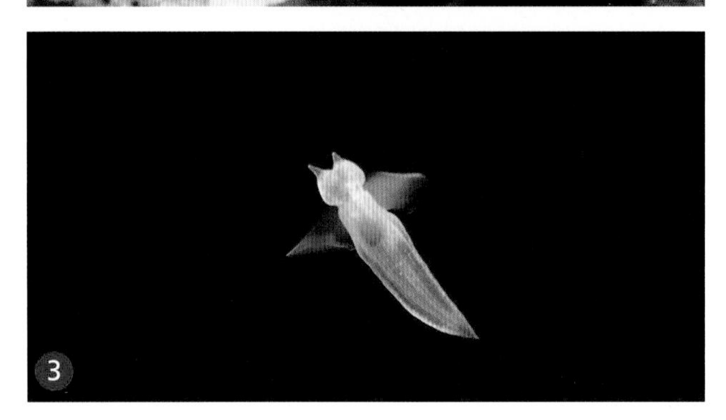

1 A queen conch snail on the sea bed in the Bahamas

2 A cone snail showing the eyes on the ends of the tentacles

3 A sea butterfly snail swimming through the water

Snail anatomy and locomotion

Sea snails are found in all areas of the sea, including coastal shores, estuaries, and deep oceans. They have a similar anatomy to land snails — a shell, a soft slimy body, and a head with a mouth and tentacles. However, the shells of sea snails are more diverse in design. Sea snails eat a great variety of foods such as algae, plants, and decaying organic matter. They are so good at eating up tiny specks of food that they are sometimes used in aquariums to keep tanks clean!

Figure 1 shows the shell, head, and two tentacles of a queen conch snail on the sea floor. At the end of each tentacle there is an eye and a smaller sensory tentacle for added sensing. Figure 2 shows a cone snail with two tentacles pointing outward. Snails move slowly along surfaces by means of expanding and contracting muscles in their body. Some snails like the sea butterfly can even swim through the water (figure 3).

A tough shell

The shell of a sea snail is secreted by a part of the snail called the mantle and is grown gradually in layers. It is made of typically around 95% calcium carbonate (chalk) and 5% organic material. The addition of the organic material makes the shell up to 1,000 times tougher than if it was just calcium carbonate. Engineers are researching shell material to make better composite materials for engineering applications.[3]

3 David Williamson and Bill Proud, "The Conch Shell as a Model for Tougher Composites," *Int. J. Mater. Eng. Innovat.*, 2011, 2, 149–164.

4 The Florida horse conch is one of the largest sea snails. **5** A queen conch shell **6** A basketful of cone shells for sale **7** The Venus comb

Beautiful variety of shells

Sea snails come in an enormous variety of shapes and sizes. Some are as small as a pin head while others can be as large as 2 feet long. The Florida horse conch shown in figure 4 is one of the largest sea snails in the world. The queen conch, which lives in tropical parts of the western Atlantic, has one of the most beautiful shells. It often has a flared thick outer lip and a characteristic pink-colored aperture (opening) as shown in figure 5.

Cone shells are found in warm tropical waters all around the world. They have a great diversity of patterns and come in hundreds of different species (figure 6). The speckled patterns are produced when the snail secretes pigments periodically. The Venus comb is a predatory snail that lives in the Indian and Pacific Oceans (figure 7).

Humans and sea snail shells

Many species of sea snail are edible and are eaten raw in salads or cooked for use in soups and curries. Conch shells can be used as wind instruments by cutting a hole at one end and blowing into the shell like a horn. Conch shells are sometimes used as decoration.

The beauty and variety of sea shells provides further evidence of the creation of beauty for beauty's sake.

Cuttlefish

Cuttlefish have an incredible ability to alter the color and patterning of their skin to match their surroundings and produce effective camouflage; they can even produce moving patterns.

Despite their name, cuttlefish are not fish but mollusks with similar features to squid and octopuses. They are found mostly in warm shallow water (figure 1) and their diet includes small fish, shrimp, and crabs.

Male cuttlefish have eight arms and two tentacles that encircle the mouth as shown in figure 2. The arms are covered in suckers to aid gripping things. Females are similar except they have only six arms.

Swimming by jet propulsion and undulation

Cuttlefish can move rapidly by jet propulsion. They suck in water then squirt it out suddenly, which pushes them in the opposite direction. The water squeezes though a funnel which can be angled so that the cuttlefish move in a particular direction. Each cuttlefish has a horizontal fin along the side of its body that undulates to help propel it at slower speeds (figure 3). The undulating fin gives a high degree of maneuverability so that the cuttlefish can navigate around rocks and coral reefs. Engineers have used this undulating fin as inspiration for small underwater robots.[4]

Unique internal shell

Cuttlefish have a unique internal shell structure called the cuttlebone. It acts as a buoyancy device with gas-filled and water-filled chambers. By changing the amount of gas in the cuttlebone, the buoyancy is adjusted and this helps the cuttlefish to maintain a similar density to sea water. Cuttlebone is used by jewelers as a mold because it is easy to carve and can withstand high temperatures. Also it

1 A cuttlefish in shallow waters in the Mediterranean

2 Cuttlefish have arms and tentacles which encircle the mouth.

3 European common cuttlefish swimming using the side fin

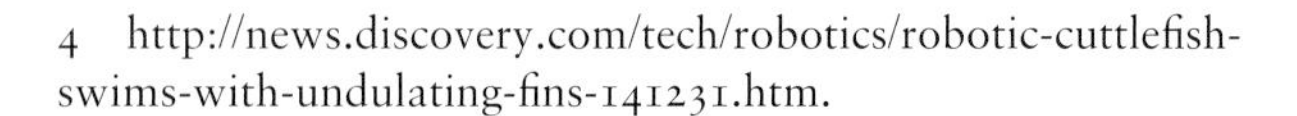

4 http://news.discovery.com/tech/robotics/robotic-cuttlefish-swims-with-undulating-fins-141231.htm.

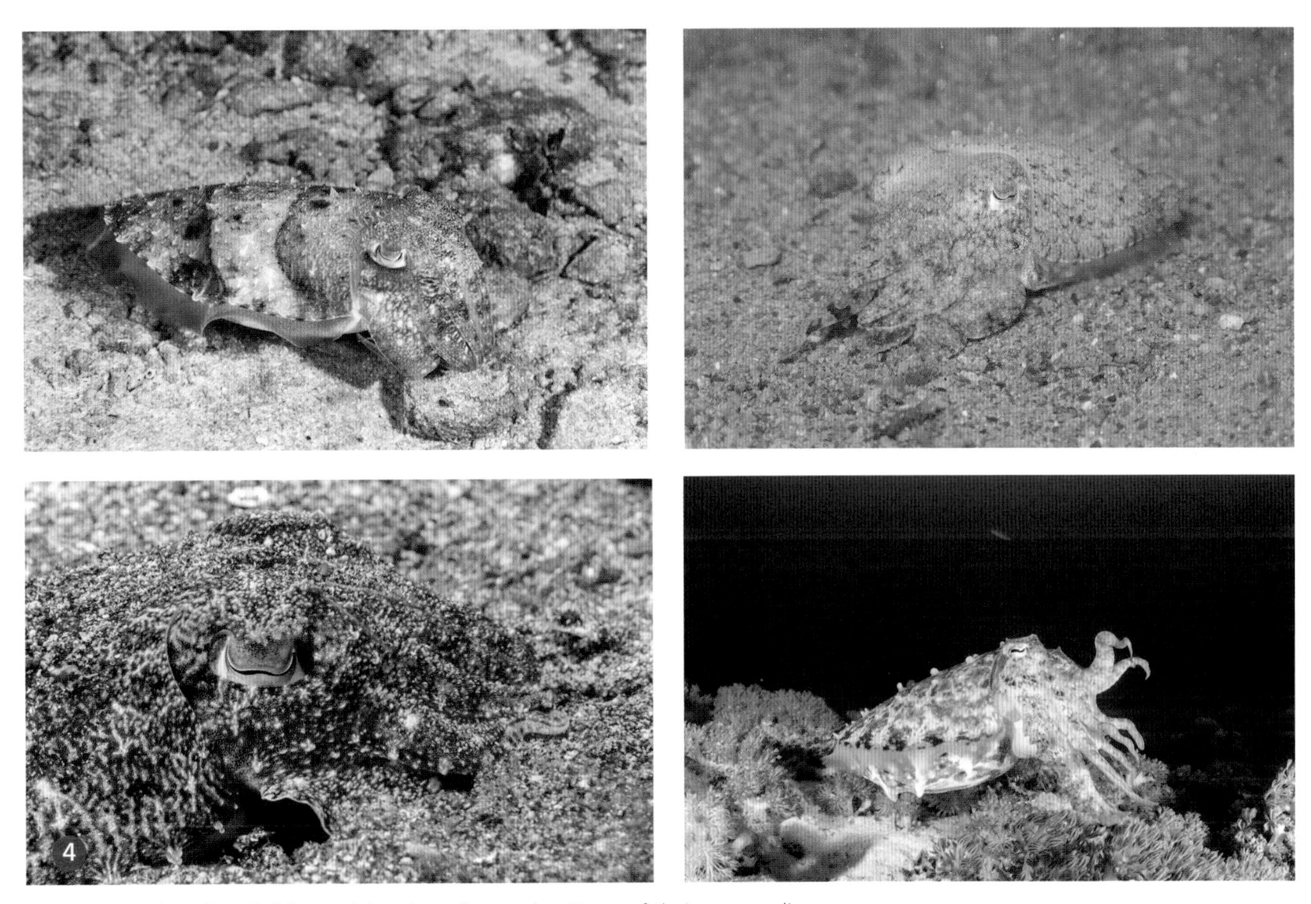

4 Four examples of cuttlefish matching the colors and patterns of their surroundings

is sometimes given to domesticated and pet birds as a source of calcium.

A visual display unit

Cuttlefish have a truly amazing ability to change the color and patterning of their skin rapidly in order to be camouflaged or warn off potential predators. Figure 4 shows different examples of cuttlefish creating colors and patterns to match their surroundings. Scientists do not fully understand how a cuttlefish can assess its surroundings and then produce such exact coloring. However, they do know that cuttlefish have very sophisticated cells in their skin called *chromatophores* which contain pigment and reflect light. There are up to 200 of these specialized cells per square millimeter, enabling fine patterns to be produced.

A typical color-changing cell has a central sac containing granules of pigment. The sac is surrounded by a series of muscles, and when the cell is ready to change color, the brain sends a signal to the muscles and they contract. The contracting muscles make the central sacs expand, creating a change in color. Cuttlefish can even produce moving patterns such as bands of light. The cells expand and contract in synchronization to produce waves of light. This can sometimes mesmerize potential prey, like crabs for example, and stop them from escaping.

Bio-inspiration

Engineers at the University of Bristol have designed artificial materials that mimic the color-changing skin of animals like cuttlefish for use in "smart clothing" and camouflage applications.[5]

5 Jonathan Rossiter, Bryan Yap and Andrew Conn, "Biomimetic Chromatophores for Camouflage and Soft Active Surfaces," *Bioinspiration & Biomimetics*. Published online May 2, 2012.

Jellyfish

Jellyfish are some of the most numerous creatures in the seas and an important part of the marine ecosystem. They have extraordinary variety and beauty.

Despite their name, jellyfish are not fish but a type of plankton made from gelatinous (jelly-like) material. Jellyfish are generally bell-shaped with tentacles and consist of up to 98% water. They do not need a respiratory system as their skin is so thin that oxygen diffuses through it. Jellyfish have no brains, no bones, no heart, and no eyes!

Blooming beauty

When huge numbers of jellyfish appear together it is called a "bloom" or a "swarm" (figures 1 and 2). In some areas of the world, millions of jellyfish can swarm together. In 2015, the southern coasts of the United Kingdom experienced particularly large swarms of jellyfish.

Smart swimming

Jellyfish can sense the ocean current and swim against it if they are drifting. Researchers have calculated that jellyfish are some of the most energy efficient swimmers in the seas. They move through the water by radially expanding and contracting their bell-shaped bodies to push water behind them. When muscles contract, the circular bell of the jellyfish rapidly becomes a smaller diameter and this pushes out a vortex of rotating fluid that gives a forward push to the jellyfish. The bell is so elastic that it expands back to its original diameter without any effort by the jellyfish. The mechanism, called "passive energy recapture," works well at low speeds and allows the animal to travel much farther each swimming cycle.

Smart eating

Jellyfish eat most edible things they come across. They can survive for a long time without food because they simply shrink when they stop eating. In contrast, they expand in size when they eat plenty of food. Sometimes they can even regrow parts bitten off by predators. When creatures like plankton come into contact with jellyfish tentacles, they are stung and then brought to the mouth.

A key part of the ecosystem

Jellyfish are a very important part of the marine ecosystem and food chain. They provide food for many creatures such as turtles (figure 3) and fish

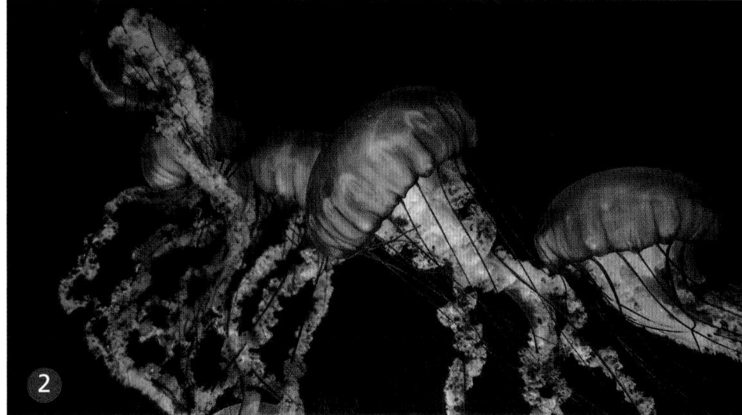

1 Spotted jellyfish blooming **2** Pacific sea nettle jellyfish blooming

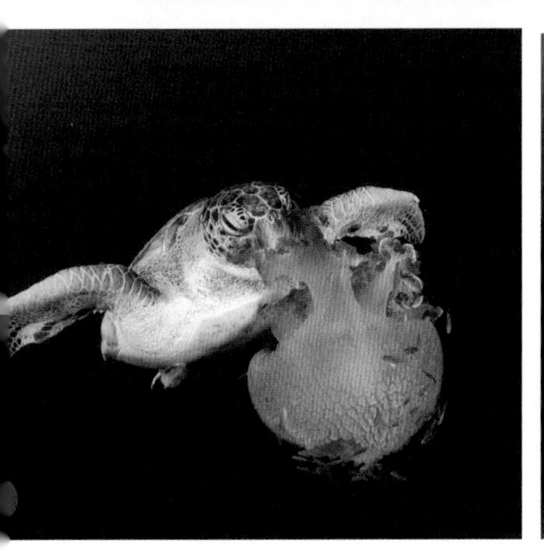

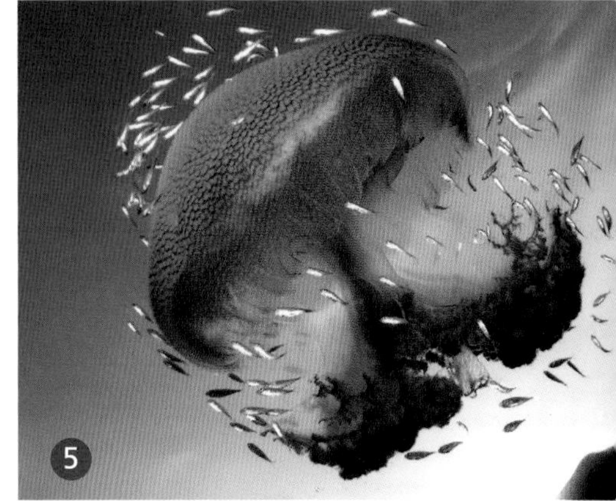

3 Green sea turtle eating a jellyfish **4** Rabbitfish eating a jellyfish **5** Small fish often swim with jellyfish for protection.

(figure 4). However, apart from providing food, jellyfish help other creatures in different ways. Some young fish are immune to the poison of jellyfish and hide in their tentacles for protection (figure 5). This provides a safe environment because predators avoid poisonous jellyfish.

Beautiful variety

Jellyfish exhibit an amazing variety of shapes and colors, some of which are shown in figure 6. The number of tentacles on a jellyfish can vary between two and hundreds, and the design of these tentacles can vary between short and long, straight and wavy, fat and thin. Jellyfish are translucent but they contain pigments which give colors such as violet, red, blue, and pink and they contain micro-organisms that produce other colors like green. They also have bioluminescence and can emit multi-colored flashes.

The incredible beauty and variety of jellyfish is evidence that they are not primitive creatures but wonders of God's creation.

6 Four beautiful jellyfish

Penguins

Penguins use their wings for swimming under water with great speed and agility. They also have a remarkable ability to nest in freezing places like the Antarctic.

There are at least 17 different species of penguins and these live almost entirely in the southern hemisphere from the Antarctic up to the Galapagos Islands. Penguins spend up to 75% of their time in the sea but they breed on land where they live in colonies (figure 1).

Emperor penguins are the largest, being about 3.3 feet tall and weighing around 88 lbs. Little penguins are the smallest at around 1.3 feet tall and weighing only 2.2 lbs. Penguins have a diet which includes fish, krill, and squid.

Underwater flyers

Penguins are said to be flightless birds, as if their wings have lost functionality, but in reality their wings are brilliantly designed to enable them to swim under water with great speed and agility (figure 2). When swimming, they have a streamlined shape like a fish. They use their tails to help steer in water and balance on land. Penguins can swim at over 20 mph and dive to depths of over 1,640 feet. They can also jump up onto banks by swimming at high speed and shooting upward out of the water (figure 3). Penguins are able to survive in sea water because they have a gland (called the supraorbital gland) that removes excess salt from their blood.

1 King penguin colonies can contain hundreds of thousands of penguins.

2 African penguin making a turn under water

3 Gentoo penguin jumping out of the water

4 Emperor penguins huddling in the cold *(Photo from C. Gilbert, S. Blanc, Y. Le Maho, and A. Ancel,. "Energy Saving Processes in Huddling Emperor Penguins: from Experiments to Theory, J. Exp. Biol. 211, 1–8, 2008; see http://jeb.biologists.org/content/211/1/i.1)* **5** King penguin incubating an egg **6** King penguin feeding a chick

Designed for the cold

Penguins have a layer of blubbery fat and many feathers that overlap and trap air for insulation. In addition, they have a layer of waterproof feathers. Penguins preen themselves to keep their feathers in good condition which is essential for their survival as they need to be able to keep warm in icy water and on exposed freezing areas of land. The waterproof feathers are kept in good condition by the penguin spreading oil over them which is produced by a special oil gland (see also page 54). Adult penguins molt each year and have to stay out of the water until all their new feathers are fully grown. Penguins have small extremities — like the bill, flippers, and feet — so less blood goes to those areas and less heat is lost. Emperor Penguins are able to recapture 80% of heat escaping in their breath through a complex heat exchange system in their nasal passages.

Penguins form huddles to conserve body heat and protect them from the wind (figure 4). When huddling in the cold, they rotate their position so that no penguin has to spend all its time on the outside. It has been estimated that huddling can reduce heat loss by 50%.

Special nurseries

Parents have a featherless patch of skin on their belly called a brood patch for warming the egg or chick when it rests on the feet of the parent (figures 5 and 6). The incubation temperature for most penguins is approximately 96.5°F, but it is a little lower for the larger species. Emperor penguins can maintain an incubation temperature of 87.8°F in an environment that is -76°F!

The parents of a chick work together by taking turns to look after the chick while the other goes to sea to catch fish. They can swallow their food whole while swimming and are able to regurgitate it to feed their young. If chicks are left on their own, they are found by their parents by voice recognition. Even in a large colony, parents and chicks have an amazing ability to find each other. One advantage of living in colonies is that there are crèches (figure 7) and babysitters! Penguins are a marvel of creation and defy evolutionary explanations.

7 A nursery of King Penguin chicks

04 Birds and Flight — Bones and Muscles

Birds are a unique created kind. Their uniqueness is defined by their powerful muscles and lightweight bone structure. Though other creatures have the ability to fly, only birds use wings made of feathers.

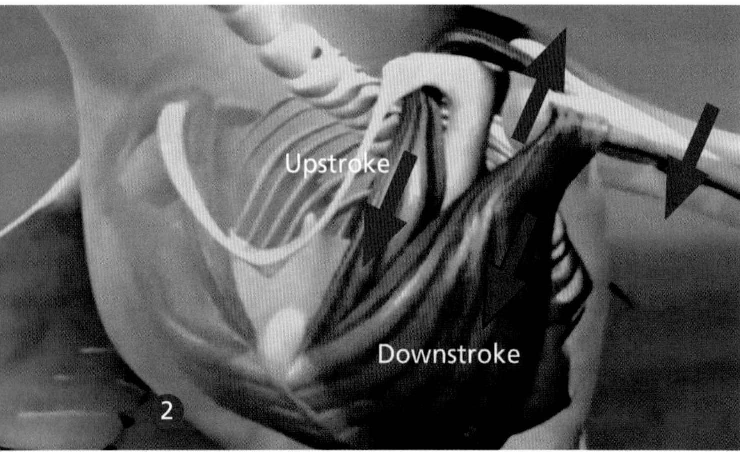

1 White-tailed eagle *(Photo by Colin Mitchell with permission)*

2 Flight muscles of a bird — the supracoracoideus muscle is underneath the larger pectoralis major muscle, and both are connected to the sternum (breastbone), so they pull in tandem.

Most of us appreciate the flight of birds. Their ability to master the air gracefully (figure 1) and often noiselessly is very impressive. Birds also have a unique organ used by many for singing. They have:

1. A light bone structure.
2. Powerful muscles which enable the strong downbeat of the wings.
3. Special muscles (figure 2), unique to birds, which enable the upstroke of the wings.
4. Aerodynamically shaped wings.
5. Precisely positioned feathers such that each feather is shaped for that particular location.
6. A preening gland near the tail. This provides oil, enabling the bird to care for its feathers to keep their flying surfaces in good condition.
7. The ability to turn the head 180 degrees to reach the preening gland and preen the tail.
8. The ability to breathe using a continuous flow mechanism so that the air continues without reversal. This is a very efficient system, getting the best throughput of oxygen for the least weight of lung.
9. The ability to sing two notes at once with a special organ called the syrinx. Musical ability is greatly enhanced as a result.

Bone structure

The bone structure (figure 3) of a bird is exceptionally light. Apart from some sea-diving birds (for example the diver, called the loon in the United States and Canada, and the puffin), their bones are hollow with cross members to give extra strength. In place of marrow, most birds have air cavities connected to their breathing system (p. 56 and 57).

3 Bone structure of a bird showing rib cage — note the articulated ribs

4 Simple pulley. The bird's flight muscles are mimicked by the simple pulley in which the effort is in pulling down rather than up.

Birds have articulated ribs, which means that the rib cage can move substantially more than in mammals. Such a movement is essential for their unique breathing system.

Extra powerful muscles for the wing upstroke

Mammals and reptiles have an outer muscle *(pectoralis major)* that produces the powerful downstroke of the *humerus* (forearm bone). However, birds have another muscle called the *supracoracoideus* which is behind the larger *pectoralis major* muscle (see figure 2). The ligament of this muscle is literally threaded round the *scapula* (part of the shoulder structure) and then over the coracoid (a short projection in the shoulder which connects the shoulder joint to the keeled sternum — the breastbone). The ligament from the *supracoracoideus* then finally comes down and connects to the sternum. When the supracoracoideus muscle contracts, it pulls the *humerus* bone (supporting the wing) upward.

This is a clever design because both muscles are themselves attached to the base of the *sternum*, such that they act in tandem with each other. As the outer *pectoralis* muscle pulls down for the downstroke, the upstroke is provided by the *supracoracoideus* acting as a pulley, lifting the *humerus* up. The threading of the ligament around the *scapula* is called the "coracoid process" and such an arrangement is found in all birds and not in reptiles — which are supposed to be the ancestors of birds according to evolution.

Humans have long known that a rope and pulley system (figure 4) makes lifting a weight easier. Instead of pulling upward, the effort is translated into pulling downward. That is precisely what is happening with the special *supracoracoideus* muscle in a bird; it acts as a pulley. There is also a second design principle whereby both muscles attached to the breast bone operate in tandem *at the same point* for the downstroke and the upstroke of the bird's wing. A third design principle is that the weight of the two muscles is at the base of the bird as it flies, giving it stability.

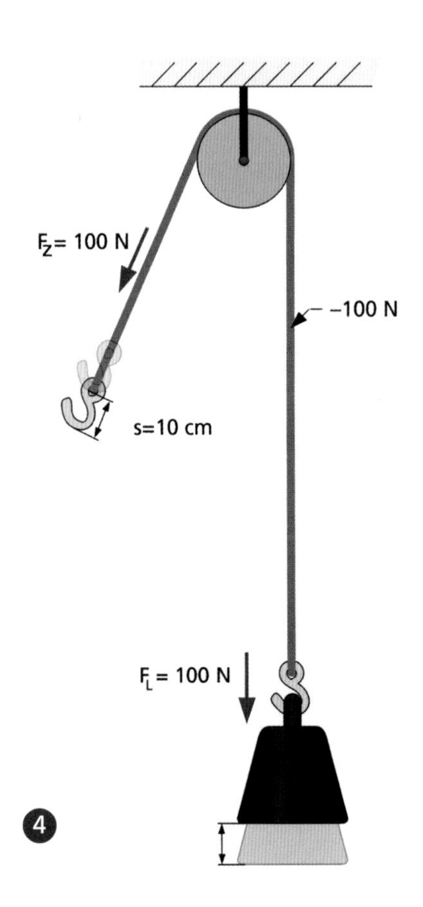

Wings and Feathers

The uniqueness of birds is particularly illustrated by the remarkable design of their feathers. There are no other creatures which have them.

Specially shaped wings for aerodynamic lift

For years the Wright brothers had carefully examined the camber of the wings of birds. They realized that wings produce lift both by the angle of the wing to the overall air flow and by the wing shape. They copied the shape of bird wings (figure 1) before making the momentous first controlled man-made flight in their Wright Flyer in 1903 (figure 2).

The shape of the wing in birds is all-important and is built up by amassing greater thickness of feathers near the leading edge so that the wing is cambered like an aerofoil.

Lift can be produced with any shape at an angle to the main air stream, but for good control the cross section of the wing matters. In birds, the shape of the wing is constantly changed during flight (see figure 1). This is not readily achieved with man-made machines, though there have been some models — like the F-111 (figure 3), Tornado (figure 4), and Mirage G — where this has been partially achieved, but the penalty in added weight for such designs has been very expensive. The Wright brothers actually did control the shape of the wings in flight by a system of pulleys attached to the wing surfaces (made of fabric) so that in many ways they were far ahead in their thinking.

Precisely positioned feathers

Feathers are not all the same. They have to be placed in the right position for each different type of function of that feather — their length and shape are precisely determined (see figure 5). Those at the tail are each symmetrical, and together the tail feathers (called *retrices*) create a fan shape used in stability for landing and in-flight maneuvers.

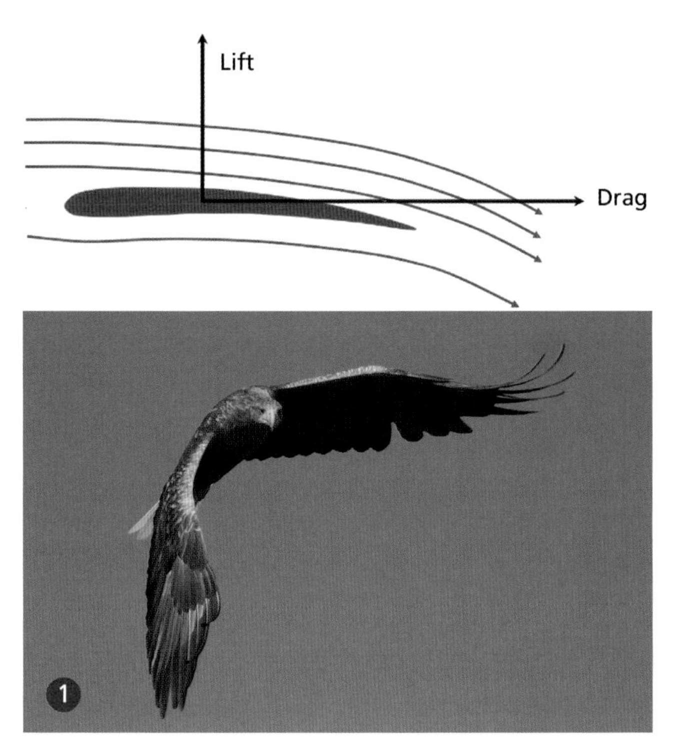

1 Wings produce lift by both angle and shape. *(Photo of white-tailed eagle by Andy McIntosh)*

Feathers on the extremity of the wing are the primary flight feathers and are heavily asymmetrical. That is, they are very narrow on one side of the central main shaft (called the *rachis*) and wide on the other side. Lying closer to the bird's body are the secondary and tertiary wing feathers; these are more symmetric. Positioned at the leading edge of the wing are the smaller covert feathers. These are primary, greater, median, and finally lesser coverts. The overlapping nature of all the feathers is such that the thickening to the leading edge of the wing resembles the curved (aerofoil) shape which increases lift control.

2 Wilbur and Orville Wright copied the wings of birds to make the first heavier-than-air controlled flight from Kitty Hawk, North Carolina, on December 17, 1903. Orville is flying the machine and Wilbur is running alongside.

3 The Swing-wing F-111

4 The Swing-wing Tornado

5 The different types of feathers on a bird must all be in exactly the right place for controlled flight.

Intricate sliding mechanism in feathers

However, there is much more to feathers than meets the eye. Feathers are made out of keratin which is a protein also used to make hair and fingernails. There are differences in the exact type of keratin used. Feather keratin occurs in a "β-sheet" configuration (many sheets pleated together), which differs from the α-helices that occur in keratins in mammals.[1] The β keratin of bird feathers is rather like a stretched spring in consistency. Some suggest that dinosaurs are the precursors to birds, using the fact that scales of reptiles are also made of β keratin. However, in fact, the stretched spring consistency in the β-sheet keratin of feathers is a major difference to that of reptiles which is not in the stretched spring form. Thus, there are substantial obstacles to transforming the β keratin of reptiles into that of birds.[2] Also, the feather grows from a follicle which is totally different from the way a scale grows.

Under the microscope (figure 6) the sophistication and detail involved in the barb system of the flight feathers of birds becomes evident. From the central *rachis* are barbs which form the vanes of the feather (see figure 6).

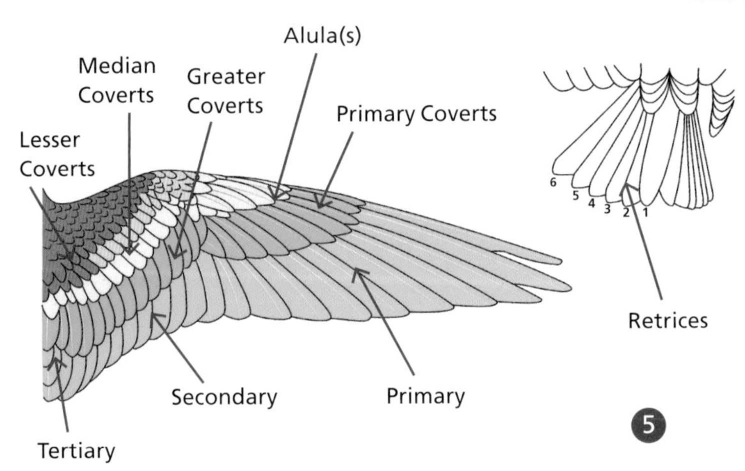

1 R.H.C. Bonser, L. Saker, and G. Jeronimidis, "Toughness Anisotropy in Feather Keratin," *Journal of Materials Science*, 2004, 39: 2895–2896. Also see website of Richard Bonser at http://www.rdg.ac.uk/biomim/personal/richard/keratin.htm, which has a useful summary of the material properties of feathers.

2 N.J. Alexander, "Comparison of α and β Keratin in Reptiles," *Cell and Tissue Research*, Volume 110, Number 2, June 1970, 153–165.

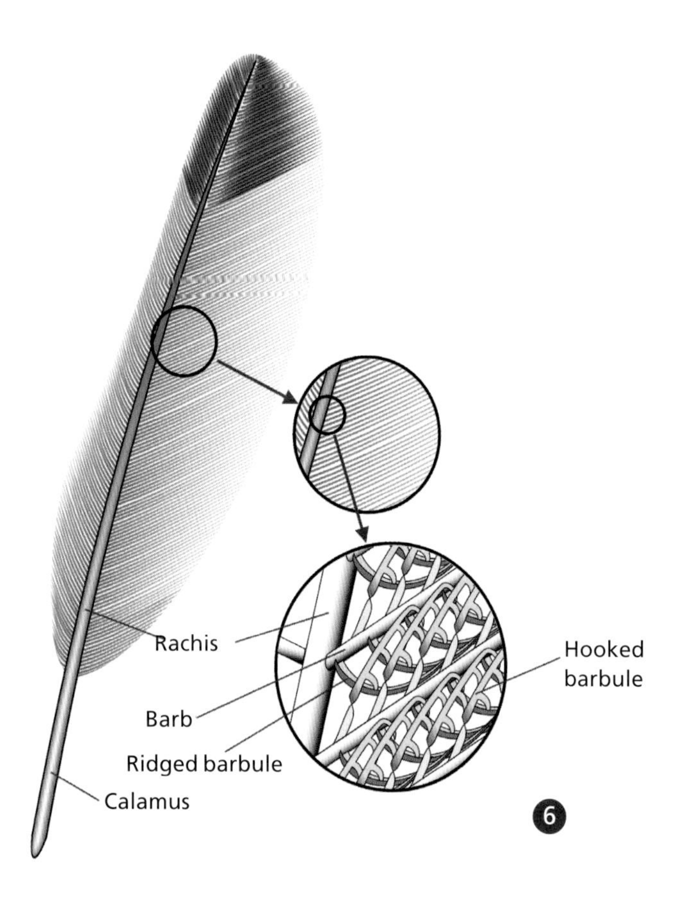

6 Feather diagram showing the microscopic hooked and ridged barbules coming from either side of the barbs.

Barbules come out from either side of each barb. They are only visible under the microscope (figure 7), but have a structure which is essential for feathers to work as aerodynamic surfaces. The barbules in one direction are ridge-like, while the barbules in the opposing direction have hooks. Consequently, the hooks of each barbule in one direction grip the ridges of the opposing barbules and slide over them. Thus, adjacent barbs are held together by a microscopic sliding mechanism between each barb.

Preening gland for oiling the sliding joint

In order to keep its feathers in trim, a bird must have a special preening gland which is positioned at the base of its spine (see figure 8). This gland provides oil which the bird picks up on its beak and then spreads over its feathers. In particular, the sliding joint of each barbule is then oiled sufficiently for the feather surface to adapt its shape under different wing movements. Without this the feathers become frayed and eventually useless.

All birds have a preening gland and all birds must have the ability to twist their necks 180 degrees (and usually more) so that they can get the oil from the preening gland, and also to reach all their feathers. The bird must have the gland and the ability to twist its neck in order to get to the gland and all its feathers. Owls (see figure 9) can turn their heads 270 degrees in an instant.

Each detail of the feathers and oil gland is evidence of irreducible complexity. Nothing works unless all is in place. This is the mark of sophisticated engineering.

Feathers for beauty

As explained on pages 136–139, the colors of feathers on birds are due to pigments and optical effects. For parrots, the only pigments are red and yellow, and the other colors such as green (see figure 10) are produced by a combination of melanin pigment and feather structure. The iridescent (structural) colors of the display feathers of peacocks are evidently there for beauty, and are produced by thin film interference where the light reflects differently depending on the transparent film surfaces through which the light has come. Some have suggested that coloration is linked to sexual selection, so that in the case of the peacock, the peahen will mate with the best performing male. Actually mating is little affected by these performances, and it has been known for a peahen to get behind the peacock and peck at him while he has his tail up!

The beauty of bird coloration is more to do with what humans appreciate, and the golden ratio 1.618 (see figure 11) is known to be linked to that which is pleasing and aesthetically beautiful (see page 145). Added beauty and overdesign is a strong evidence of God's fingerprint on creation.

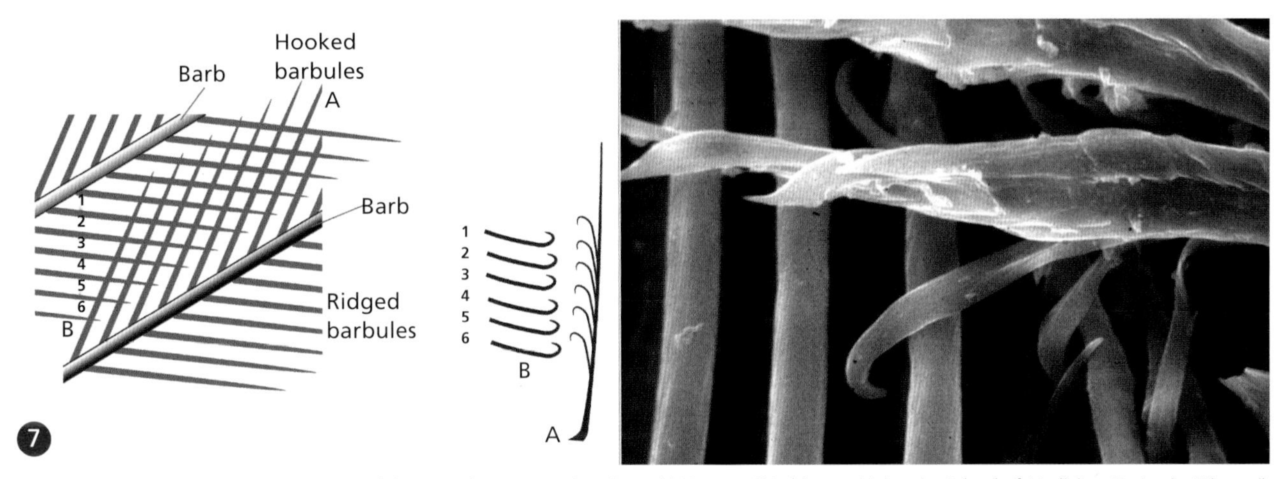

(Photograph courtesy of Prof. David Menton, Washington University School of Medicine, St. Louis, Missouri)

7 Under the microscope, the detail of the hook and ridge sliding connector system for flight feather barbules is stunning.

8 A mute swan preening **9** Owls, such as this barn owl, can turn their heads 270 degrees.

10 Male Australian king parrot — Alisterus scapularis. *(Photo by Toby Hudson, Wikimedia)* **11** The eye of a peacock feather. The ratio of the two lengths is the golden ratio.

Breathing System

Birds have a unique breathing system using air sacs linked to a continuous flow lung. They have no diaphragm and instead each bird moves its breast bone in and out.

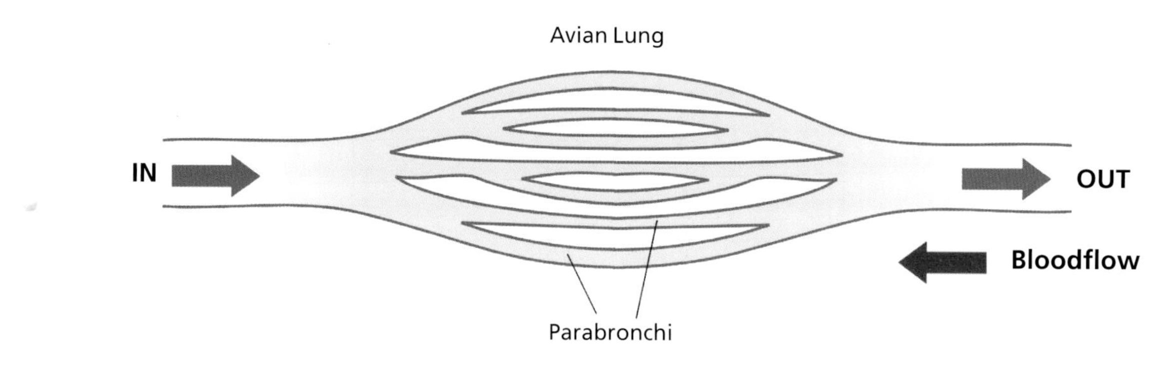

1 The parallel counterflow lung in bird respiration

The remarkable design of a bird's breathing system

Mammals and most reptiles have a diaphragm and breathe so that at each inhalation the air comes to a halt before being exhaled at the second part of the cycle.

Birds have an entirely different system where the air is in continuous transit through the lung (figure 1). Instead of a network of ever smaller tubes called *alveoli* in a mammalian lung and most reptiles (like an inverted tree of tubes which spread into the tissue of blood vessels), the bronchial passage in birds splits into parallel tubes known as *parabronchi* so that air is continuously flowing through the lung. This continuous flow system is essential for a bird. Its breathing is much faster than that of a mammal or land-based reptile because the energy expenditure (metabolic rate) is much greater.

It is well known in engineering that a mass exchanger is at its most efficient when a counterflow system is in use. This means that the most efficient system for the transfer of gas across a membrane from one fluid to another is when those fluids flow in opposite directions.[3] So in the case of the bird lung, this means that the blood goes in the opposite direction to the air. It is this principle that is being used in the respiratory system of the bird lung, which is effectively a series of counterflow mass exchangers between the blood flowing in one direction and the air flowing in the other. Consequently, this uses a minimal amount of tissue to achieve a given rate of gas exchange and the energy involved in inhalation and exhalation is also minimized.

In addition to this special air-flow system, unlike mammals and reptiles, a bird has air sacs (even inside its bones) where air is stored as it goes round the continuous system (see figure 2). On the first inhalation the air goes to a rear air sac, then during that exhalation an earlier packet of air comes out of the throat (that is the *trachea*) while the air it has

3 F.P. Incropera, D.P. DeWitt, D.L. Bergman, and A.S. Lavine, *Fundamentals of Heat and Mass Transfer*, Sixth Edition (London, New York: Wiley, 2006).

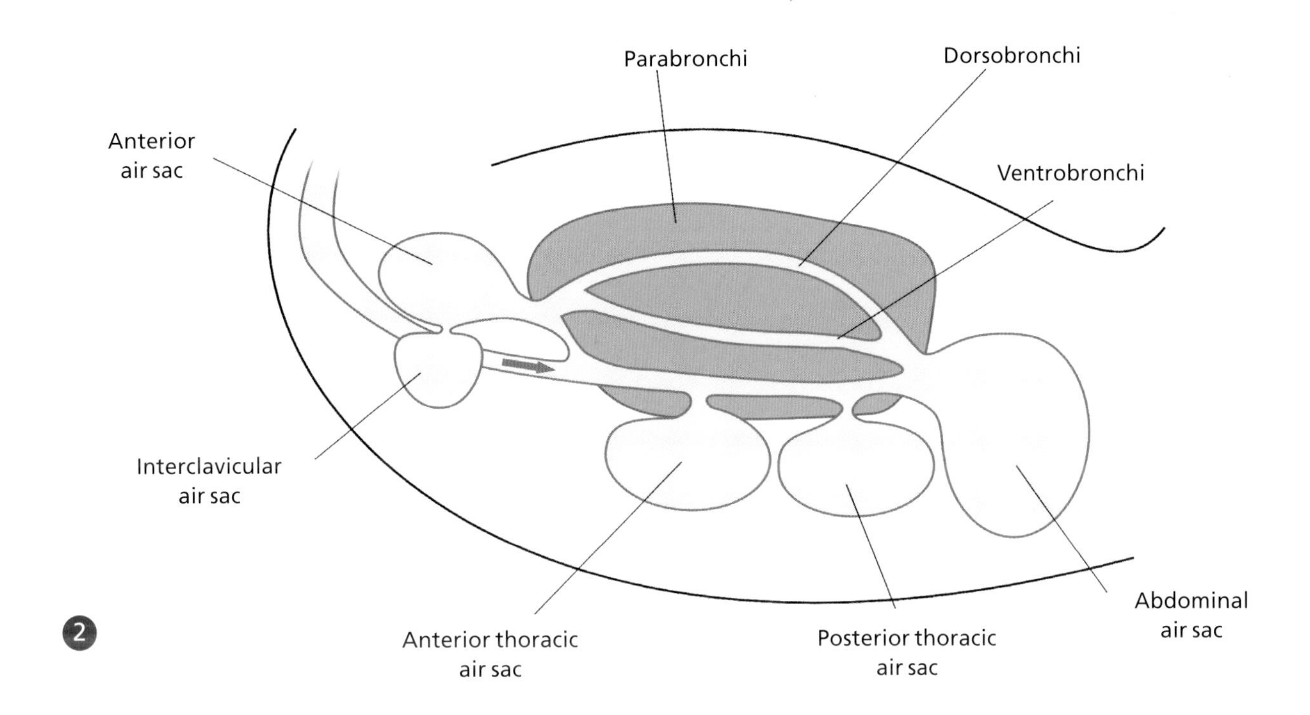

just breathed in goes through the lung (always in the same direction). When the bird breathes in again, the air we are following moves to a front air sac and then, when it breathes out the second time, this air we are following finally comes out of its mouth. This enables the continuous mechanism of breathing so that a high amount of respiration is possible.

2 Diagram of a bird's respiration system, showing the many air sacs which keep the lung supplied with much oxygen. The lung of a bird has the air going through in only one direction. Some have tried to suggest that alligators and crocodiles have a similar parabronchial breathing system but even if this were proved, the big difference is that the whole system is operated by the sternum (breastbone) and a bird has no diaphragm!

Birds do not have a diaphragm

Unlike mammals and reptiles, a bird has no diaphragm! Some believe that the supposed evolution of birds from reptiles is supported by the fact that the alligator lung has some features similar to the *parabronchi* tubes in the lungs of birds and thus it is argued that they have a similar unidirectional breathing mechanism like birds. However, unlike birds, these reptiles move air into their lungs by contracting the diaphragm, and operating rib (intercostal) muscles, in the same way as we and all mammals do.

In contrast, a bird contracts muscles controlling the *sternum* (breastbone) to move air round the circulatory system. When it breathes in, the *sternum* moves forward and downward with a combined movement of vertebral ribs so that the internal pressure is lowered and the posterior (rear) and anterior (front) air sacs inflate in succession. This counterflow unidirectional breathing system of the bird, coupled with a moveable sternum is unique.

Music from the Syrinx

Humans, cetaceans (whales and dolphins), and birds are the three created kinds that have the ability to sing. Birds do this using a syrinx organ which can even allow some to sing two notes at once.

The syrinx

Nearly all perching birds *(passerines)* are also singers and represent the largest group of birds. All birds have an amazing structure called the *syrinx* which is located where the *trachea* forks into the lungs (see figure 1). The muscles modulate the sound by changing the tension of the membranes and the bronchial openings. There are membranes in both halves of the fork in the windpipe, and these can be made to produce two sounds at once; this gives the exquisite quality to bird song. Their notes and songs are often not just simple changes in frequency but involve complicated calls made by a number of notes and timings.

Some birds such as parrots, crows, and myna birds can mimic speech. But the bird which is by far the best at mimicking other sounds is the Australian lyrebird (figure 2). This bird has even been known to imitate chain saws![4]

Why birds sing

The evolutionary explanation of the phenomenon of bird song is often limited to the idea that the bird with the best song gains the fittest mating partner. The actual evidence suggests that each bird identifies itself at an early age by the song it learns from its parents and therefore variations simply establish identity so that in mating, bonds are created by the calls and songs.

As with the colorful beauty of feathers there is clearly overdesign in bird song. It is far more complex than would be expected for mere functionality. Many bird songs enhance our lives with beauty which is linked to our own ability immediately to distinguish sounds of different frequencies.

The quality of bird song

While some bird calls are not noted for beauty but are simply raucous sounds, such as the crow or magpie, others have distinctive and delicate songs. Bird song is analyzed by using sonograms

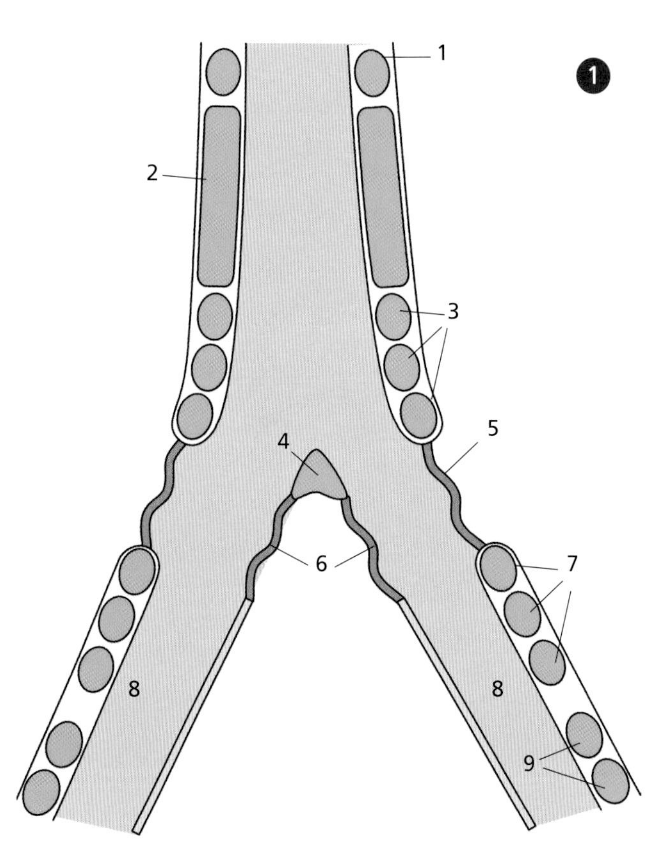

1 The bird's syrinx. The numbers 5 and 6 are the walls of twin (tympaniformis) membranes that vibrate as the air is passed over them. The pessulus (4) also vibrates and with twin bronchial tubes (8) close to the vibrating membranes, two notes at once are possible. The other numbers refer to muscles used to control the sounds made.

4 This is shown in the series by David Attenborough entitled "Life of Birds."

2 The lyrebird has an amazing mimicking ability.

3 A sonogram measuring the frequency of sound against time in seconds for a robin

such as that illustrated in figure 3 for the European robin. All bird species can be recognized by the general characteristics of their song or call. Some individual birds have a particular style within those characteristics so that the song of an individual bird is unique (see p. 140–141 for further discussion on the structure of bird song). Well-known examples of English birds with their distinctive songs or calls are robins, chaffinches, wrens, blackbirds, and sparrows. Some birds even sing at night, like the nightingale, and it is often the case that robins and song thrushes will sing well into the darkness of evening. Although songbirds do not generally keep to a diatonic or chromatic scale, their songs do involve repeated strains which are not random. Phrases will be emitted many times before moving to a different strain.

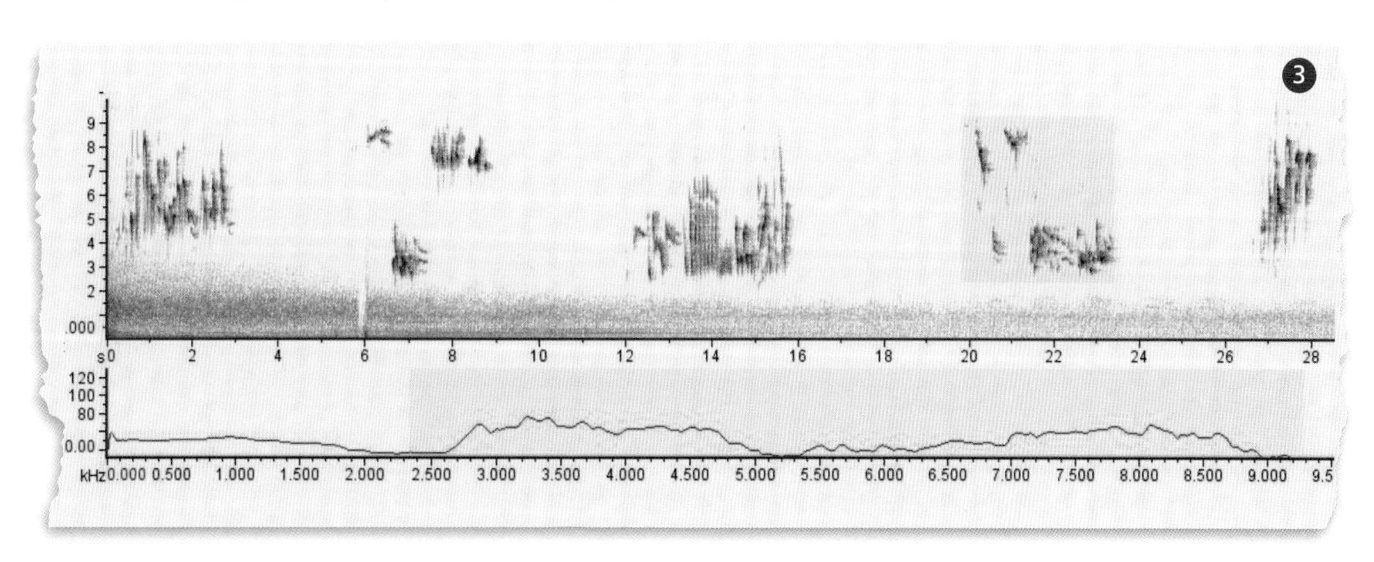

Distance Gliding — The Albatross

The soaring of an albatross over vast distances using updraft from waves over the open ocean is a demonstration of great gliding skill.

Specially designed wings

There are 21 species in the albatross family, and these range across the Southern Ocean, Pacific Ocean, and South Atlantic.

The albatross (figure 1) has very long wings for the same reason that gliders (figure 2) have long wingspans. This design feature is to reduce the aerodynamic drag, called "induced" drag, that all finite length wings will inevitably have. The vortices (spinning air) that spill from the edge of the wings into the slipstream are really lift that is lost at the wingtips. Increasing the span of the wing reduces this effect.

With a wingspan of 11.5 feet, the wandering albatross is a very powerful glider and can stay on the wing for days on end. For every beat of its huge wingspan, a wandering albatross can cover a distance of 72 feet. By using gliding principles of flying into the wind to gain height, it can glide for hours before beating its wings, and thus stays aloft for days, weeks, and even months. It will also fly low over the ocean which reduces drag further, and will gain extra lift using the updraft from ocean waves as the swell of the sea moves the air currents immediately above the sea.

Special salt gland

Because they drink a lot of salt water, like all sea birds the albatross has a special salt gland. This gland is situated above the nasal passage and helps to desalinate their bodies. Humans cannot drink salt water, and many have died of thirst when on an open ocean. For humans to desalinate water it takes a special process and involves reverse osmosis

1 The shy albatross with a wingspan over 8 feet breeds off Australia and New Zealand's subantarctic islands and ranges over the Southern Ocean.

2 A glider with its large wingspan flies by the same principle as the albatross, and reduces the aerodynamic drag which is always the penalty of lift.

3 The wandering albatross (distinguished from other albatross species by its pink bill) has a wingspan up to a gigantic 11.5 feet and breeds on islands such as South Georgia in the Southern Antarctic seas (with wings curved down to further reduce drag).

(a chemical membrane treatment separating water on one side of the membrane from the salt solution on the other side), heating to draw off the steam and then recondensing the steam into water. Complex chemical plants are used and are not cheap to run. In contrast, all sea birds use special glands to cause salt in the blood to be drawn out chemically into two ducts alongside the bill where the salt solution is excreted out of the bird.

Distance Record

Wandering albatross (figure 3) spend most of their lives soaring across the ocean. They can easily fly distances of well over 300 miles in a day. In November 2013, a wandering albatross was reported to have traveled 10,000 miles in a single journey of 30 days from the Kerguelen Islands in the Southern Indian Ocean to Southern Scotland.[5]

The ability of the bird to fly such large distances has only recently been understood. With a strong surface wind such as is common across the oceans, and without flapping their wings, albatross are similar to a fixed wing glider. However, they use a soaring technique to gain height with almost no energy loss, as their wings are kept straight out. They angle their whole body into the wind and rise vertically, then they swing round and descend downwind in a long arc; they then turn again into the wind with a slightly greater forward velocity than the first time. In this way they gradually increase in height and soar across vast distances with no wing motion needed at all. Though glider pilots use this technique to soar across country, the albatross achieves much longer distances right across the globe. It has been calculated that an albatross uses more energy landing and taking off than in its long distance flight!

5 See http://www.dailymail.co.uk/sciencetech/article-2509211/Mystery-wandering-albatross-travels-10-000-miles-single-journey-WITHOUT-flapping-wings-solved.html.

Aerobatics — Hummingbirds

The aerobatics of hummingbirds is astounding. They generally beat their wings at 50 to 60 times each second but have been known to reach an astonishing 150 times per second. They can fly forward, sideways, and even backward.

How do hummingbirds fly?

The flight of hummingbirds involves complicated aerodynamics. In every wing beat cycle there is a twisting and untwisting of the wings (figures 1, 2). The wing tip traces a figure eight motion as shown in the diagram in figure 3. The normal downstroke (see stages 6–15 of figure 3) leads at the end of this motion into a twisting of the wing (stages 16–20). The now twisted wing on what would normally be an upstroke, is not bent as with other birds, but is still being used as an aerodynamic surface. On this second part of the cycle, the twisted wing is now using the upper surface to beat effectively another downstroke (stages 21–35). Finally, at the end of the complete cycle, the wing untwists (stages 1–5) and another downstroke occurs.

In figure 2 one can see clearly the twisting of the wings that is involved in hummingbird flight. The bone structure is specially designed so that the ball and socket joint at the end of the shoulder bone is able to swivel to a much greater degree than in other birds. This enables the humerus bone to twist, along with the radius and ulna. Recent studies have shown that the wrist bone can also twist.[6]

Use of Vortices

When they beat their wings, all birds generate vortices (spinning air) from the tips of the wings. These create a downdraft behind their wings but an updraft at the extremity of their wings. Because the hummingbird is beating its wings so fast, it is able to pick up the motion in the air from the previous beat to its advantage.

Figure 4 shows the vortex rings created as a series of "doughnuts" from each wing cycle. These are then manipulated by the hummingbird as it darts back and forth. Figure 5 shows the path of the wing tip in forward, upward, and backward motion. Sideways movement is generated by a slightly different motion (wing beat frequency / angle of

6 T.L. Hedrick et al., "Morphological and Kinematic Basis of the Hummingbird Flight Stroke: Scaling of Flight Muscle Transmission Ratio," *Proc.* Roy. Soc. B. 279, 1986–1992, 2012.

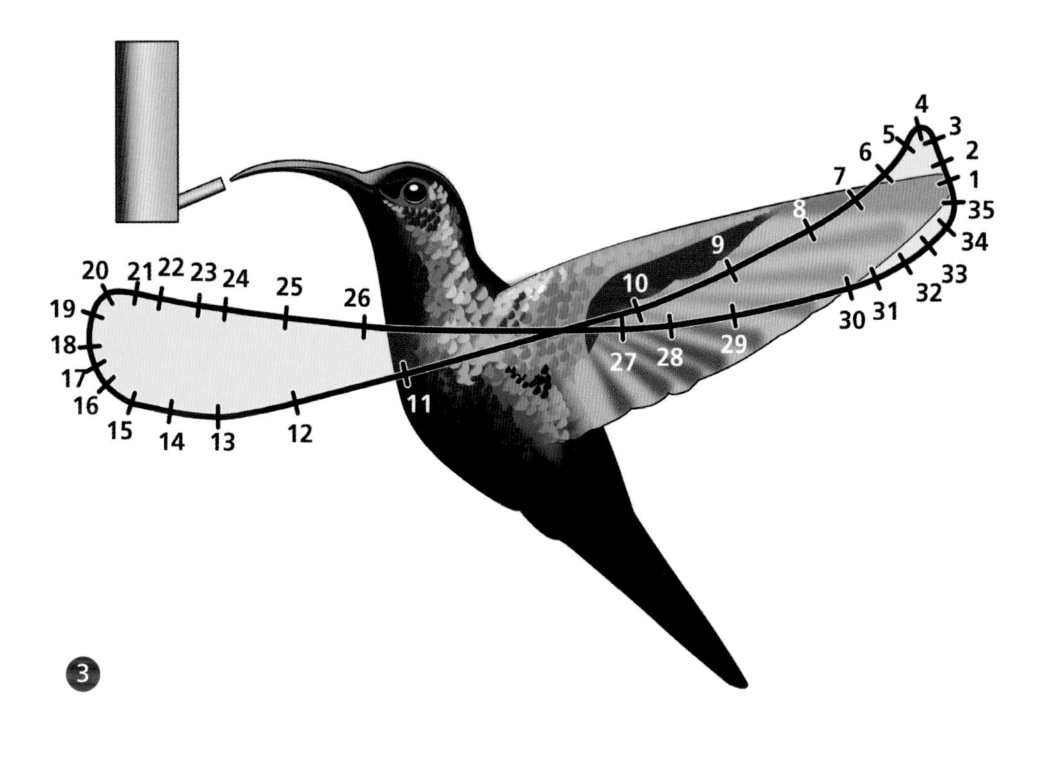

1 Anna's hummingbird

2 Blue-throated hummingbird clearly showing the twisting of the wings. *(Photo by R.W. Scott)*

3 Wing tip motion of a hummingbird

4 Hummingbirds use ring vortices created from each cycle of wingbeat. A series of "vortex doughnuts" is created from each wing cycle.

5 Hummingbird wing tip motion for forward, upward, and backward motion

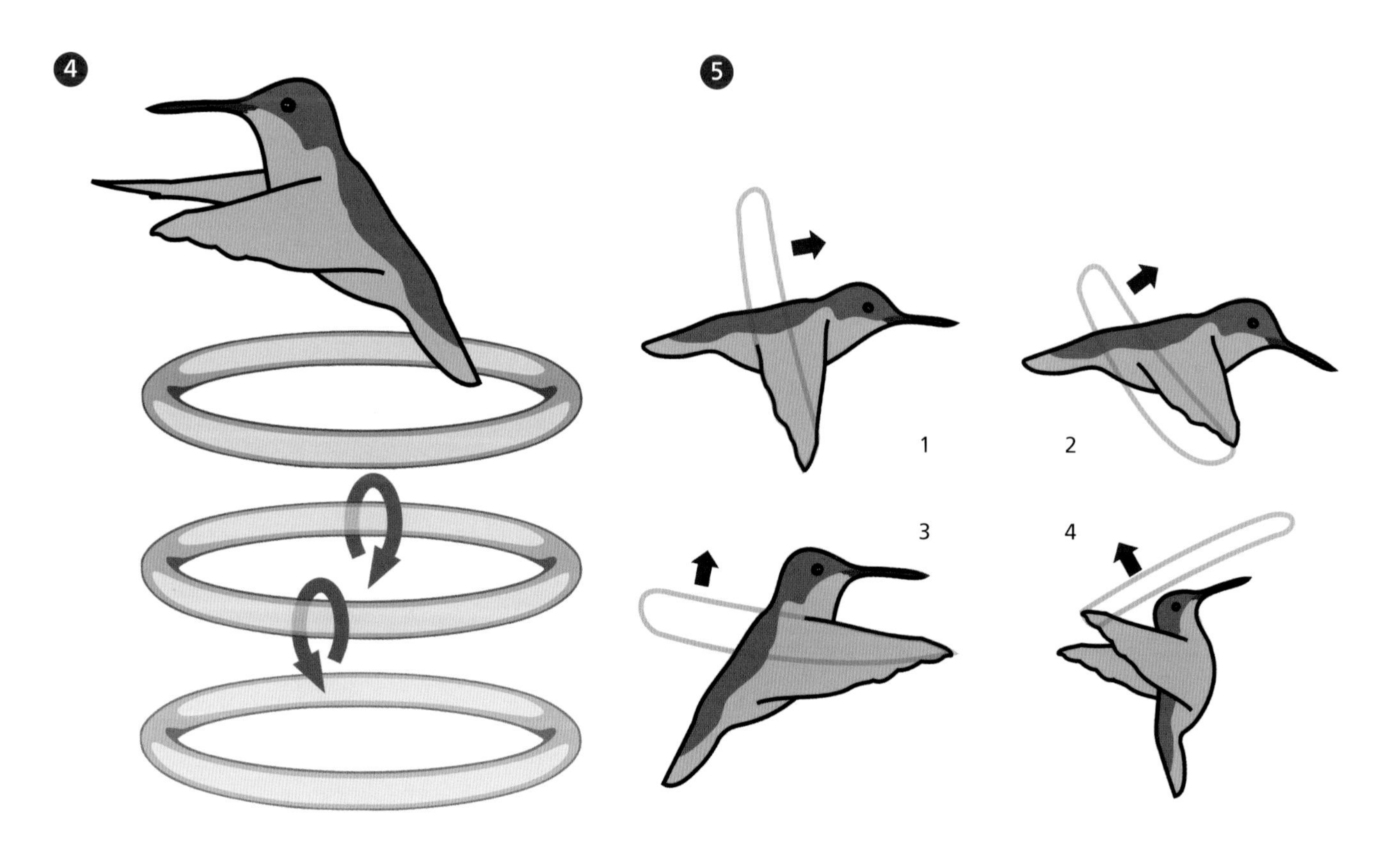

inversion of wing, etc.) of the left hand wing to the right, or vice versa.

Long beak and tongue essential for feeding

All hummingbirds feed on liquid with high energy sugars. This is usually nectar from plants or sap from trees which they sip while hovering. The long tongue can be seen protruding from the beak of the broad-billed hummingbird shown in figure 6. By having a sugar solution in the hollow of one hand, it has even been known for these marvels of the air to sometimes sip the solution with their long tongues while still hovering.

The tongue emerges from the beak at 5–10 times per second as the wings are beating 50–60 times per second. The tongue is well designed to enable the lapping up of the nectar. In cross-section, it is double-grooved and flexible so that as the tip curls up, the tongue laps up the nectar, which pours down its throat. Hummingbirds consume almost their own bodyweight in food every day, in order to survive.

These marvelous birds illustrate a well-proven principle in engineering that in multi-component systems there are many parts which are interdependent and all must work together in order for the whole system to work. If one item is missing or malfunctioning, then everything fails. For example, there are many systems in aircraft which all work together. As soon as one item is missing or malfunctioning (such as elevators or ailerons) it will be catastrophic. The same is true for hummingbirds. For instance, they need the long beak and specially designed tongue as well as the ability to hover. Having one of these abilities on its own is no advantage. All these functions are needed together.

Migration of hummingbirds

When hummingbirds are preparing to migrate long distances in the summer, the need to feed takes on an even greater urgency. They require an increase in their original body weight of 25–40%. In migration they will generally fly during the day and sleep at night, roosting in trees. They travel no higher than the treetops in these journeys and are able to obtain nectar from flowers on the way. However, some of the routes of migrating hummingbirds take them long distances over water (figure 7), and then they must feed exactly the right amount — too little and

6 The broad-billed hummingbird with its tongue protruding

7 Ruby-throated hummingbirds migrate across the Gulf of Mexico to overwinter and breed in the warmth of Mexico. They cover 450 miles in 20 hours, sometimes against a 20 mph headwind.

8 The smallest hummingbird is the bee hummingbird, found only in Cuba. At 2–2.5 inches, it is barely larger than a bee and can perch on the end of a pencil or a twig. Its wings beat even faster than larger relatives at 80 cycles per second.

9 Almost as small as the bee hummingbird is the bumblebee hummingbird, found in Mexico and which is 2.8–3 inches. *(Photo used by permission and under licence from Pete Morris, Birdquest)*

they will not make it, too much and they will be too heavy for the journey.

Some hummingbirds, such as the rufous hummingbird, migrate as far north as Alaska, and cross the Mojave Desert both there and back. Some fall out of the sky during this arduous trek, but most survive.

Sleeping torpor

If the temperature drops at night, hummingbirds have the ability to go into a torpor, which is a deep sleep similar to the hibernation of some animals. Hummingbirds lack insulating downy feathers as they are generally in warm environments and downy feathers would cause them to overheat. They are very sensitive to temperature, so at night as the temperature drops they lower their internal thermostat and become hypothermic (losing heat faster than they can gain it and so cooling the body). No human can cool their own core body temperature; if this is forced on us we will die when the body temperature drops by more than a degree or two below 98.6° Fahrenheit — as all mountain climbers and arctic explorers are aware.

When the hummingbird becomes hypothermic, it actually turns its own body temperature down, and lowers its rate of producing energy (metabolic rate) by as much as an astonishing 95%. It is barely alive and could be thought of as dead since no motion in the lungs can be detected and it will not wake when touched. For it to recover normality out of torpor can take as long as 20 minutes. During this period, heart and breathing rates increase and the bird vibrates its wing muscles (shivering); this generates heat, warms the blood, and quickly changes the body temperature by several degrees each minute. Even the smallest hummingbirds (figures 8, 9) go into torpor overnight.

There appears to be nothing other than the bird's circadian clock (internal day/night clock) to explain its awakening out of torpor each morning between one and two hours before dawn, ready for another daytime feeding frenzy — a truly astonishing design feature!

Power — The Eagle

The flight of the eagle leaves no doubt that this is the king of the bird kingdom. Its size, strength of flight, vision over great distances, and fearsome ability to hunt outstrips all other birds of prey.

Wingspan

The wings of an eagle are impressive. Bald eagles and golden eagles have wingspans of approximately 5.9–7.5 feet, while sea eagles are larger at 8.2 feet and stand 3.3 feet tall. The largest eagle is the wedge-tailed eagle *(Aquila audax)* whose wingspan reaches 9.3 feet.

Winglets

Any wing, when it produces lift, has the effect of making vortices in the air. When these vortices slip off the end of the wing, the energy put into producing lift is lost and contributes to aerodynamic drag. This part is called "induced" drag.

In order to reduce this, birds have some amazing devices that they use to make their wings more efficient. These are feathers positioned at the wing tips which splay out in a visible fan (figure 1) for high lift /low speed maneuvers. These "winglets" powerfully reduce the effect of induced drag and are found on many birds of prey — raptors and owls — but also other large birds such as cranes.

When an aircraft is in a high lift configuration (flaps down at take off and landing) there is very high induced drag and therefore greater expense of fuel. Winglets (figure 2) reduce this drag under these

(Photo by Colin Mitchell with permission)

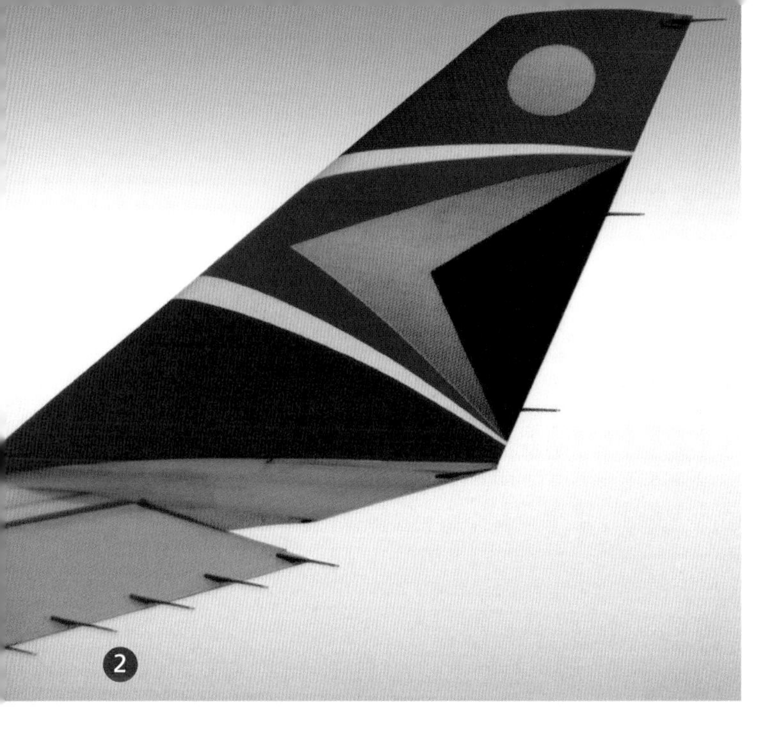

1 White-tailed eagle fishing on the Isle of Mull in Scotland. Notice the winglets on the end of the powerful main wings.

2 Right-hand side winglet on a South African Airways Airbus A340–300. *(Photo by Andy McIntosh)*

3 Like all raptors, the bald eagle has very sharp eyesight. Their forward-facing eyes enable binocular vision so that they can readily judge distance, which gives them great ability in hunting.

low speed conditions and so make airports less noisy and reduce pollution. But the idea came from bird wings in the first place!

Eyesight

Binocular vision and the fovea in the eye

Eagles and all raptors have very sharp eyesight, enabling them to focus effectively on great detail at long distances (figure 3). There is also a very sensitive part of the eye called the *fovea*. This is where the receptors (called rods and cones) in the retina are very closely packed and give greater sensitivity in vision. Human eyes have a *fovea* with approximately 200,000 receptors per square millimeter. Raptor eyes have far more, with hawk and eagle eyes reaching up to 1 million receptors per square millimeter. An eagle in flight can reputedly sight a rabbit two miles away. Each eye has two fovea which are connected with a ribbon-shaped region which acts as a third *fovea*.

Pecten

All birds have a separate organ in their eyes called a *pecten,* which is not fully understood. It is made of folded tissue which projects from the retina, and it is believed that it may shade the retina from dazzling light and aid in detecting moving objects.

Cornea

Human eyes have the ability to change the shape of the lens to focus on something nearby. Not only do eagle eyes have that same ability, but the outer clear part around the eye (the *cornea*) also changes shape in eagles, which leads to sharper near vision as well.

Special eyelids

Birds and some other animals have a third eyelid (called the *nictitating membrane*) which is transparent and moves across the eye horizontally. See page 16 which describes the extra eyelid of the camel. This is used to protect the eye particularly in flight, and to keep it moist, while still enabling the bird to see. Humans find it difficult to see in high winds because of the inability to protect the eyes. Raptors at speed, and particularly eagles, blink rapidly using this third eyelid so that their eyes are still operating normally — even when the eagle is diving to catch its prey at speeds of 125–200 mph.

Migration

The migration of birds is still one of the world's mysteries. How they not only perform long distance travel but also navigate over empty oceans, is only slowly being understood. In the Bible, God aptly reminded Job, "Does the hawk fly by your wisdom, and spread its wings toward the south?" (Job 39:26).

Astonishing examples
Pacific golden plover

Some birds can fly very large distances, often across open ocean. To do this, man-made aircraft use sophisticated instruments to work out their position. This requires considerable flying and navigation skills. The Pacific golden plover (figure 1) migrates between Northern Canada/ Alaska and Hawaii – a distance of around 3,000 miles over part of the Pacific Ocean (figure 2). As if this was not enough, it is accomplished in two phases. The adults go first in August followed by the young making the journey on their own from Alaska in October, having never done this before!

The challenges involved are considerable. The young birds have never been to Hawaii before. Not only do the young birds need to know where to go, they also need to eat enough to make sure they can make the journey. During the trip, the plovers use up approximately half of their bodyweight and have to make sure they eat enough beforehand to survive the trip, but not too much as that would then make them overweight for the trip. Too much or too little means they will not make it. Trained pilots of conventional aircraft have to do careful calculations

1 The Pacific golden plover (Pluvialis fulva) in breeding plumage

2 The staggering distance flown by the Pacific golden plover

3 The Arctic tern (Sterna Paradisaea) flies from pole to pole and back each year!

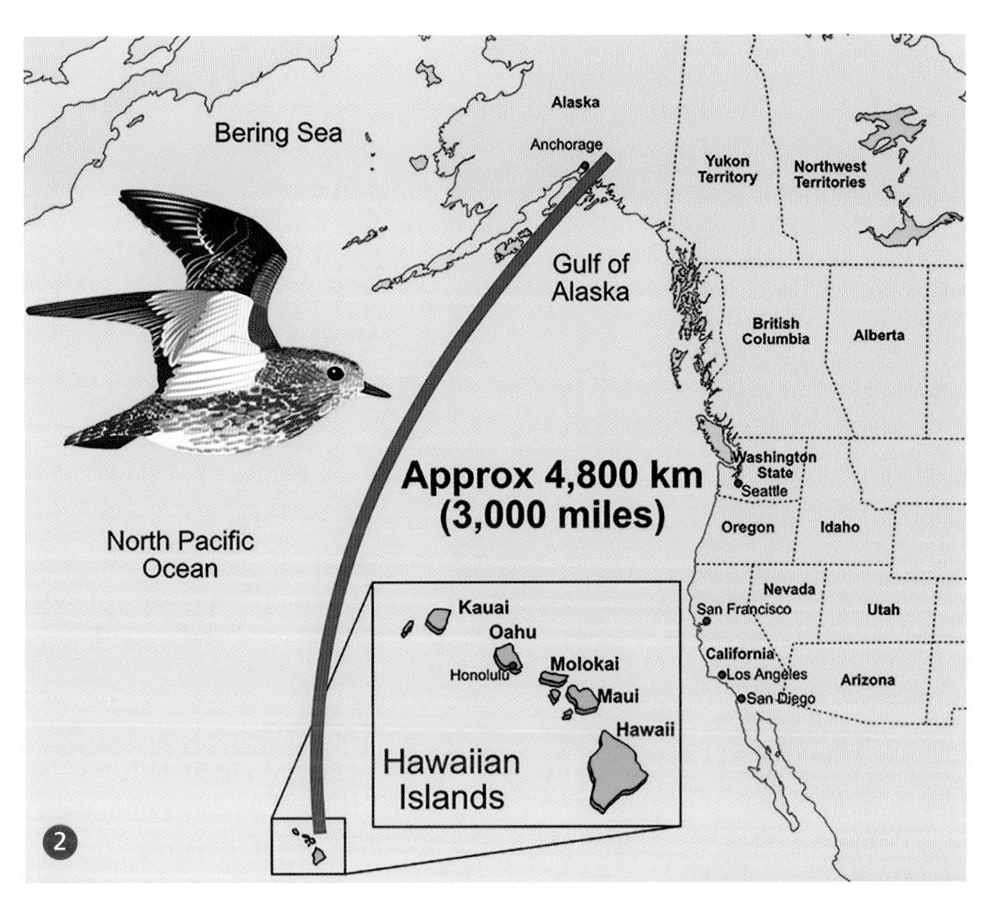

concerning fuel load and weight to assess whether they can reach a destination over open water safely. Yet these birds, which have never done it before, arrive without difficulty. This feat of engineering requires complex knowledge which cannot be obtained by a trial and error approach.

The Arctic tern

The Arctic tern (figure 3) is a seasoned traveler. It breeds in Greenland in the Northern Hemisphere, mostly within the Arctic Circle, but flies south to spend the northern winter in the Southern Hemisphere, mostly in the Antarctic ice-packs around the Weddell Sea on the shores of Antarctica. This journey to Antarctica enables the Arctic tern to enjoy the benefits of a second summer, making the most of daylight and a plentiful supply of food. Therefore, Arctic terns fly from pole to pole every six months. For birds that are only 15 inches long and with a maximum weight of 4.5 ounces, this is no mean achievement.

4 The bar-tailed godwit flies the longest non-stop flight across the Pacific of approximately 7,100 miles.

5 Canada geese in V-formation, which reduces the drag of the group

A British Antarctic Survey team from Greenland attached tiny tracking geolocators to a number of Arctic terns and discovered the astonishing fact that they flew in a huge S-shaped fashion near the coastlines of Africa or South America and that each year they covered a distance of almost 44,000 miles.[7] This made it the longest annual migration of any bird in the world. Since Arctic terns can live up to 34 years, this means that the total distance that such birds fly during a lifetime can equal three round trips to the moon — or almost 1.5 million miles!

Arctic terns breed in colonies numbering hundreds or sometimes thousands of pairs. The breeding colonies are generally very noisy, but just before migration all the birds suddenly go completely quiet and then fly up together and out to sea. Somehow there is mass communication that they are about to leave to make their epic journeys. This odd behavior is known as "dread," but the reason for it is not properly understood.

Bar-tailed godwit

The bar-tailed godwit (figure 4) takes the prize for the longest non-stop migration flight. In 2007, wildlife biologists used surgically implanted satellite transmitters to show that migrating bar-tailed godwits fly from Alaska to New Zealand without once stopping to refuel! At 7,100 miles in just over eight days, the migration was, and remains, the longest non-stop flight ever recorded.[8]

The use of Earth's magnetic field

Two important instruments are needed for travel: a map of where to go and a compass bearing. The map pinpoints where you are and a compass orients you in the right direction to move.

A number of cues have been shown to be used by migrating birds to navigate: the sun's position, smells, and even the stars. It is known also that the Earth's magnetic field is used by birds as a compass. What has been more difficult to establish is the "map" that birds evidently have. They must have this in order to pinpoint where they are at any given time while in their large migratory flights.

A clever experiment with Eurasian reed warblers has shown that the magnetic field of Earth actually provides a map as well as a compass for these birds. This experiment involved interrupting the reed

7 C. Egevang et al., "Tracking of Arctic Terns *Sterna Paradisaea* Reveals Longest Animal Migration," *Proc. Natural Academy of Sciences*, 107(5), 2078–2081, February 2010.

8 P.F. Battley et al., "Contrasting Extreme Long-distance Migration Patterns in Bar-tailed Godwits *Limosa lapponica*," *Journal of Avian Biology*, 43, 001–012, 2012.

warblers on their way northeast from the Baltic Sea to breeding grounds near St Petersburg. The birds were then moved 600 miles east, and when released they knew their location immediately and re-oriented themselves northwest. Furthermore, when the birds were taken near the Baltic Sea and placed in an environment at the same location but with a strong magnetic field which simulated the location 600 miles east, they still flew northwest.[9]

This proved that birds have a magnetic field sense — though how it works and where the sense mechanism is located has not yet been established.

When some birds (such as geese) migrate, they fly in a V-formation (figure 5). The reason for this is that the overall effort used by the group is lower due to the vortices of spinning air which come off the wingtips of each bird. These are always such that just outside and behind the wingtip the flow is upward, but just inside and behind the wingtip the flow is downward, so that a bird flying immediately behind and shifted to one side can take advantage of the upward flow from the vortices coming off the wingtip of the bird in front. Another bird can do the same behind that bird and so on. In a V-formation, the bird doing the most work is the one at the front, so one can sometimes see the lead bird move to the back and the one immediately behind taking over the increased work at the front.

It is apt to consider Jeremiah 8:7 which states "Even the stork in the heavens knows her appointed times. . . . But My people do not know the judgment of the LORD."

9 N. Chernetsov et al. "A Long Distance Avian Migrant Compensates for Longitudinal Displacement During Spring Migration," *Current Biology*, 18, 188–190, 2008.

05 Insects — Dragonflies and Damselflies

Dragonflies and damselflies are some of the most impressive fliers in the insect world, and with four wings individually controlled, they fly rather like helicopters.

1 Scarlet marsh hawk dragonfly (*Aethriamanta brevipennis*) with its very striking scarlet abdomen *(Photo by Andy McIntosh)*

Dragonflies and damselflies belong to the order Odonata[1] which covers all flying carnivorous insects. Dragonflies (figures 1, 2) are agile fast fliers, and can reach speeds of 25–35 mph. Damselflies, though also very agile, have a different mode of flying and will more often move from one hovering motion to another, somewhat like hummingbirds.

1 The technical term for dragonflies is *Anisptera* (referring to the difference between hindwings and forewings), and for damselflies the technical term is *Zygoptera* (meaning paired wings).

Life cycle

Dragonflies and damselflies have a lifespan of more than a year and there are three stages of their life cycle. They begin as eggs, which hatch in water and then quickly become nymphs. One rarely sees these, even though most of the life of the dragonfly is lived in the nymph stage. Finally they become the adult dragonfly or damselfly.

2 Meadowhawk dragonfly *(Sympetrum Vicinum) (Photo by Robert Jensen with permission)*

3 The dragonfly *(Libellula quadrimaculata)* coming out of its molt shell (called an exuvia)

Nymphs

Once the eggs hatch, the larva becomes a nymph which looks like an alien creature. Its wings are present but not grown and they are within the casing of what looks like a crusty hump hanging onto its back. This part of the life cycle can take up to four years to complete under the water, and if the nymph cycle is completed in the beginning of the wintertime, it will remain in the water until spring when the weather is warm enough for it to come out. The nymphs live in ponds or marshy areas of a stream or the calmer backwaters of rivers. They are voracious feeders and use a unique lower lip that they project to hook their prey. They may eat creatures as big as tadpoles or small fish and they have been observed to eat smaller dragonfly nymphs. They can even become cannibals and eat nymphs of their own species!

Metamorphosis

When the nymph reaches adult size, it comes out of the water never to return, and sheds its last molt (called an exuvia — see figure 3). Remarkably, the nymph which was breathing underwater using gills, now changes to breathing in air through spiracles as all land insects do. A spiracle (as figure 4 shows) is more than simply a hole in the thorax of the dragonfly or damselfly; it has a special valve so that air is sucked into the trachea behind the opening. To make sure that this valve only begins to operate once the dragonfly or damselfly nymph has left the water, a set of thin white threads is woven through these tubes and trachea in the new thorax, so that only when the exuvia is broken open are these threads pulled out of the spiracles.

The new dragonfly inhales initially with its mouth, but full respiration is achieved once the spiracles are operational. At the same time, blood is pumped through the veins of its powerful wings that then become stiffened aerodynamic surfaces. The whole process is remarkable, since not only is it effectively a new creature from the old, but it is particularly astonishing since the breathing which used gills in water is now changed skillfully to breathing in air.

King of the flying insect world

The flight of dragonflies is spectacular (figure 5) with top speeds of 35 mph. This is achieved primarily because of the individual muscle arrangement. As a consequence, dragonflies can outmaneuver most other insects and are the king of hunters in the insect world. They will catch smaller insects at speed on the wing, but sometimes smaller dragonflies themselves become the meal for larger species of dragonfly.

Occasionally dragonflies will swarm — thousands of them will mature altogether in one location and hunt insects and grubs en masse. However, dragonflies themselves are not without predators. There are birds such as kites, hobbies, wagtails,

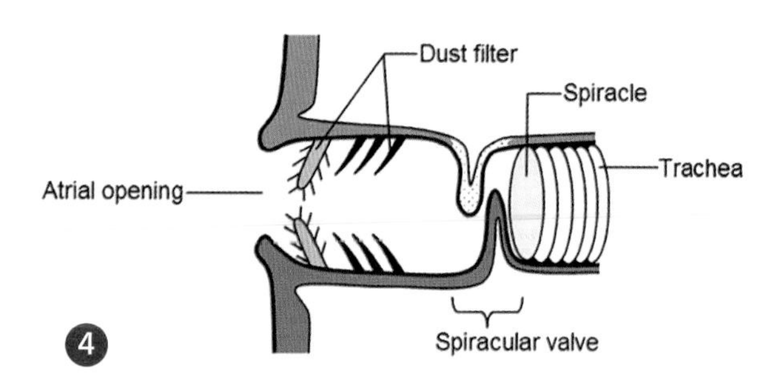

4 The spiracle through which dragonflies and all land insects breathe

5 Meadowhawk dragonfly *(Sympetrum Vicinum)* in counter-stroking flight *(Photo by Robert Jensen with permission)*

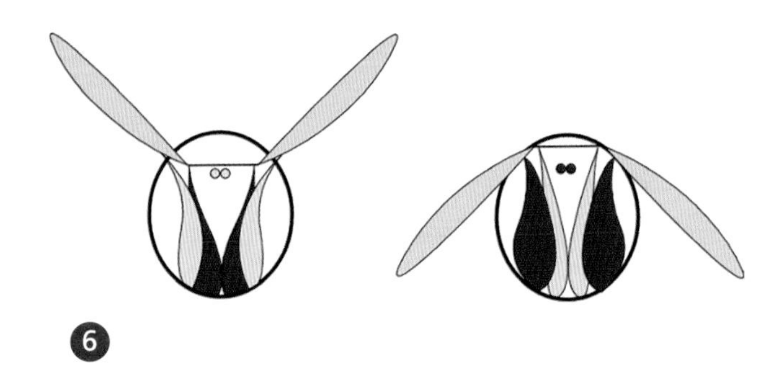

6 Direct muscular system used in the thorax of dragonflies and damselflies. Each wing has a muscle controlling the upstroke (left-hand diagram) as well as another muscle controlling the downstroke (right-hand diagram). This is similar to the twin muscle arrangement in birds, discussed on pages 50–51. See also https:// www.amentsoc.org/insects/glossary/terms/direct-flight-muscles.

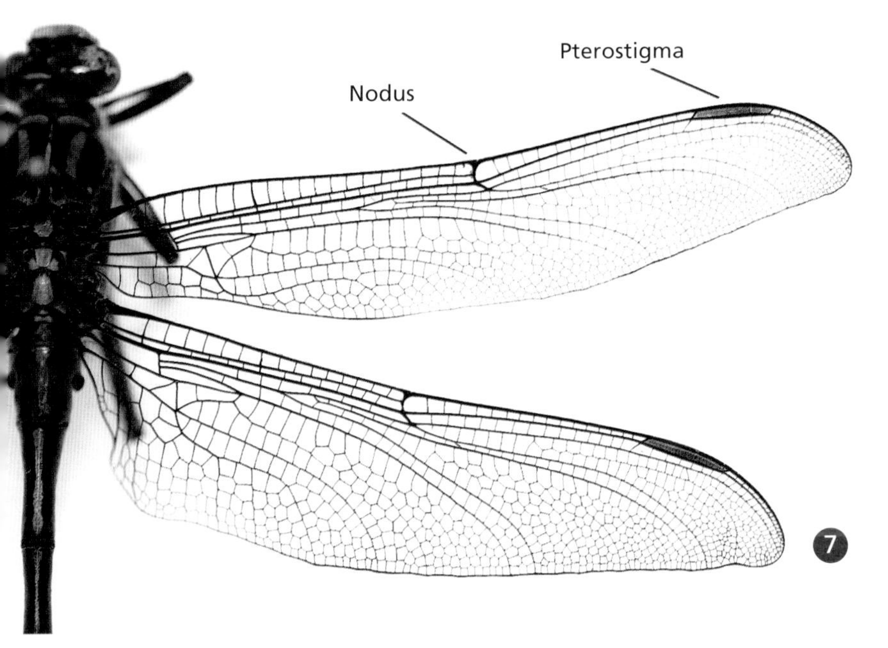

7 Wings of dragonfly showing *pterostigma*
(Photo: Brian Edwards)

and swifts that, with keen eyesight and agile flight, can actually catch this top insect sky-hunter in mid-flight!

Muscles controlling the wings

Dragonflies and damselflies have individual muscles for controlling the downward and upward motion of each of the four wings (so eight muscles in all) in powered flight (see figure 6). The only other insect to have this arrangement (called direct musculature) for powered flight is the cockroach.[2] Wasps and flies have only one muscle for each wing. A muscle at right-angles across the thorax indirectly pushes both wings back up as the cross muscle contracts. The difference with this arrangement for Odonata to that of other insects is that there is a power stroke in both directions that makes for very strong flight.

Different modes of flight

The flight muscles of dragonflies can adjust the wing beat frequency, the amplitude (that is the extent of wing beat), the phase difference between forewings and hindwings, and also the angle of attack of each of the four wings independently.

2 D.J. Borror and D.M. De Long, *An Introduction to the Study of Insects*, 3rd edition (Holt, Rinehart & Winston, 1971), p. 50–51.

The four different modes by which dragonflies use their fore and hindwings:

• Counter-stroking flight: forewings and hindwings move up and down about 180° out of phase. This pattern is generally used in cruising flight (see figure 5).

• Phased-stroking flight: the hindwings cycle about 90° before the forewings. This provides more acceleration when in complicated aerodynamic maneuvers

• Synchronized-stroking: the forewings and hindwings move in unison. This is not often used, but can be used in slower flight

• Gliding: the wings are held without beating for free gliding

Natural resonance and pterostigma

Typically, dragonfly wings beat in the range of 30 to 50 cycles per sec. (Compare the hummingbird on pages 62–63). For all Odonata, at the end of each wing and on its leading edge there are dark spots called pterostigma which are heavier than the rest of the wing (see figure 7). During flight, as they reach maximum forward flight speed, light wings (without pterostigma) would start fluttering uncontrollably — this is due to the natural vibration frequency of

8 The beautiful golden-ringed dragonfly has the appearance of a helicopter. Its exuvia is alongside. *(Photo by Brian Edwards)*

the wing being excited by aerodynamic forces, and these forces depend on the speed of the dragonfly. This is called the "resonant frequency," which depends on the mass distribution of the wing.

With pterostigma, the wings stay balanced because the mass distribution of the wing is greater at the tip, and the threshold speed — where the natural frequency for uncontrolled vibration occurs — is raised outside the range of dragonflies. So the pterostigma are precisely placed stabilizers for the wings of all odonata!

The technical term for all this is "inertial regulators."[3] This clever and detailed feature is a clear example of sophisticated design. In one species it has been shown that even though one pterostigma contributes only 0.1% to the weight of a dragonfly, it raises the critical speed for uncontrolled vibrations by 10–25%. It would be absurd to suggest that such an intricate feature had come about by random mutations since any different size and location of pterostigma would be unstable for sustained fast flight.

Wing twist and four-wing interaction

Each wing has the capacity to twist upward (supination) and downward (pronation). This is also a very important feature for gaining extra lift for each wingbeat. It is similar to the hummingbird wing motion described, on pages 62–63, and enables the dragonfly or damselfly to manipulate the vortices (spinning air) shed from each wing. This is particularly true when the wings are being used in "counter-stroking flight" (see earlier section) where the hindwings are entirely (180°) out of phase with the forewings. In this case, the hindwings pick up some of the vortices shed from the forewings, so that on the downstroke of the hindwing, the insect senses the updraft from these vortices. This greatly reduces the induced drag (that is drag due to lift) of the individual wing, and therefore the wings together are more efficient than when acting separately.

These three features (muscles, pterostigma, and wing interaction) raise a significant issue, that of irreducible complexity — nothing

3 R.A. Norberg, "The Pterostigma of Insect Wings an Inertial Regulator of Wing Pitch,", *Journal of Comparative Physiology*, Volume 81, Issue 1, p. 9–22, March 1972.

9 Female helicopter damselfly (giant damselfly — *Megaloprepus caerulatus*) in the Osa Peninsula, Costa Rica. *(Wikimedia)*

works unless everything works. It is actually the detailed knowledge of aerodynamic engineering which uncovers the intricate design features in nature's flyers.

Those who believe in evolution constantly escape behind words such as "millions of generations and chance mutations provide the genetic possibilities of change," but it is the engineering scientific detail that gives the lie to such statements. The evolutionist is committed to the à priori philosophy of evolution and is thereby compelled to ignore detail — and that is not good science.

Precision engineering requires design, and no engineer looking at Odonata wings could come to any other conclusion. Wings do not make themselves. The design of muscles to operate the wings, the design of pterostigma all in exactly the right place, the design of the four wings with a different shape for the forewings to the hindwings — all these illustrate the principle of irreducible complexity — that is (as stated earlier), nothing will work unless the whole system is operating together. This can only happen by design. A wing without the right muscle arrangement is useless and a dragonfly without pterostigma is also useless. Intricate design speaks of the Lord of creation who loves detail.

The flight of damselflies

Damselflies are similar in structure to dragonflies, though usually lighter in build. However, the wings of dragonflies are held flat and away from the body, while most damselflies hold the wings folded at rest, along or above the abdomen. Their wings are also considerably thinner and are in proportion to their comparitively much thinner abdomens.

Even though damselflies appear to have weaker flying ability than dragonflies, they are actually very able fliers and hover with precision when stalking their prey.

Nature's helicopters

Some dragonflies look like helicopters (figure 8) but damselflies are the ones nicknamed "nature's helicopters" with good reason. Though generally damselflies are smaller than dragonflies, the largest of all Odonata is in fact a giant damselfly called the helicopter damselfly (*Megaloprepus caerulatus* — see figure 9) found in Central America. This damselfly has a wingspan of 7.5 inches and it feeds on orb-weaver spiders in the forest, which it plucks from their webs.

The giant helicopter damselfly first finds a web, hovers in front of it until it locates the spider, and then flies backward and quickly darts forward again to grab the spider in its forelegs. In positioning itself, it employs individual wing movements which make it far more versatile than man-made helicopters — movements which are essential if it is to avoid being taken itself by the web. The damselfly also employs optical cunning: it carefully comes in at an angle so

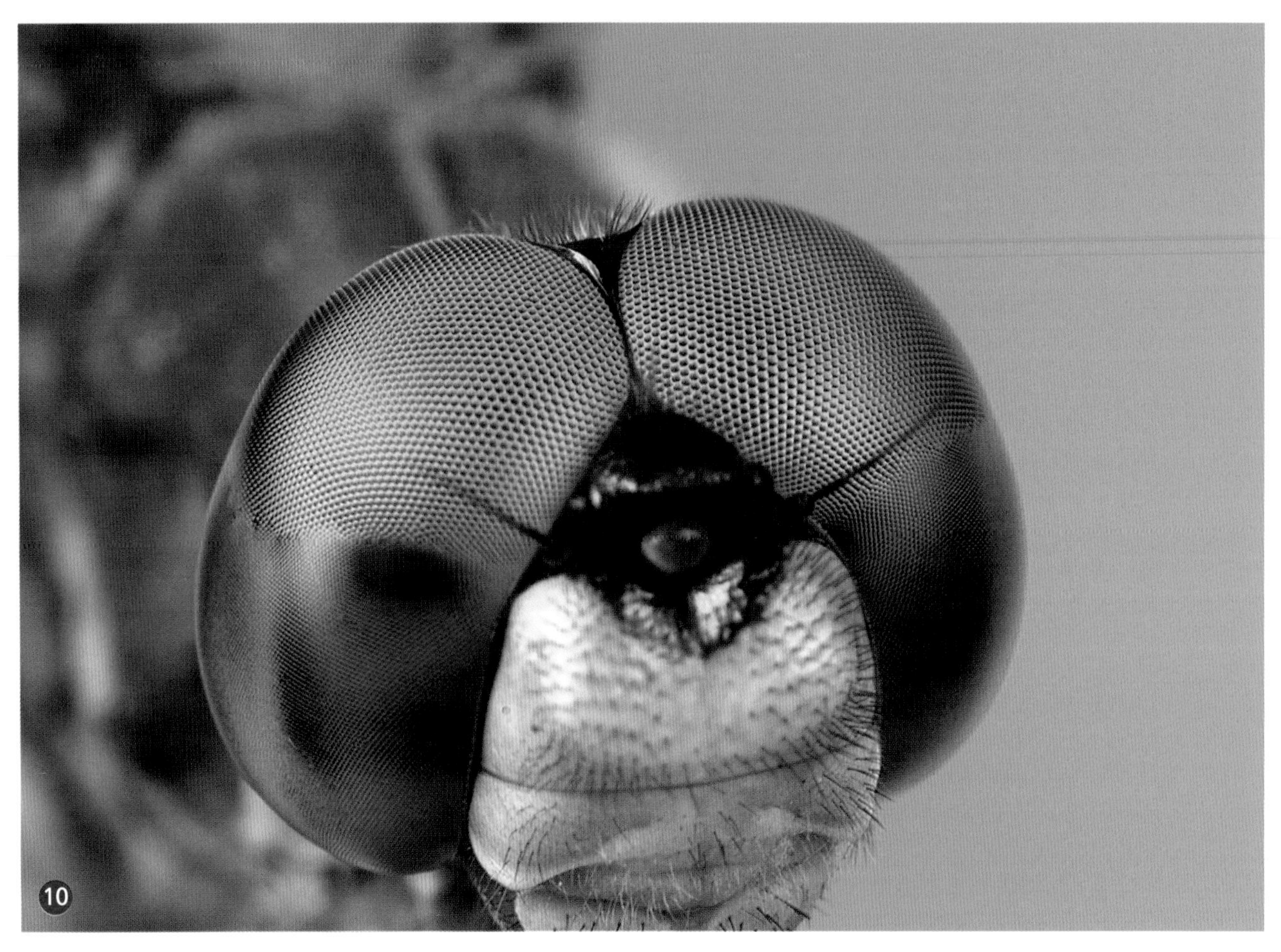

10 Close-up of the multi-lensed eye of the bluedasher dragonfly. An adult dragonfly eye has nearly 24,000 ommatidia (lenses). *(Photo by Robert Jensen with permission)*

that the spider sees no change in lateral movement as the damselfly approaches at speed to take it from the web.

The eyes of dragonflies and damselflies

The eyes of all Odonata are made of a series of thousands of individual lenses — these are called compound eyes, which all flies have. However, what is a particular feature in Odonata eyes is that they are pushed outward from their heads so that the eyes have an almost completely spherical shape that enables them to observe prey, and predator movement, in any direction.

Each compound eye can contain as many as 24,000 lenses (see figure 10), and each lens is perfectly positioned on the sphere and is connected to an individual retina behind it, with a separate nerve taking the signal to the brain. Vision in single-lens creatures is very complex, and more so for compound eyes, since the brain takes each of these thousands of images from each lens and processes them to make sense of position and direction for split-second movements.[4]

4 See http://insects.about.com/od/dragonfliesanddamselflies/a/10-Cool-Facts-About-Dragonflies.htm. Accessed March 2016.

11 Dragonfly fossil *Mesurupetala* from Solnhofen quarry, Germany — late Jurassic *(Wikimedia)*
12 Dragonfly fossil *Meganeuradae* with an approximate 2-foot wingspan found in France, and from the Carboniferous Stephanian coalfield. *(Wikimedia)*

Dragonfly fossils are found in rock, and sometimes the impressions even of the veins of the wings are accurately conserved (figures 11, 12). These fossils start in Carboniferous rock, so it is supposed that these creatures evolved 240 million years ago. However, the exquisite preservation of wing details speaks volumes as to the sudden nature of their burial. These fossils are in fact of creatures buried in the worldwide Flood, only thousands of years ago. It is worth commenting that, although supposedly from 240 million years ago, they are apparently exactly the same as those observed today — no changes!

Butterflies and Moths

Butterflies capture the imagination with their apparently haphazard flight, but they are in fact masters of the air and fly large distances.

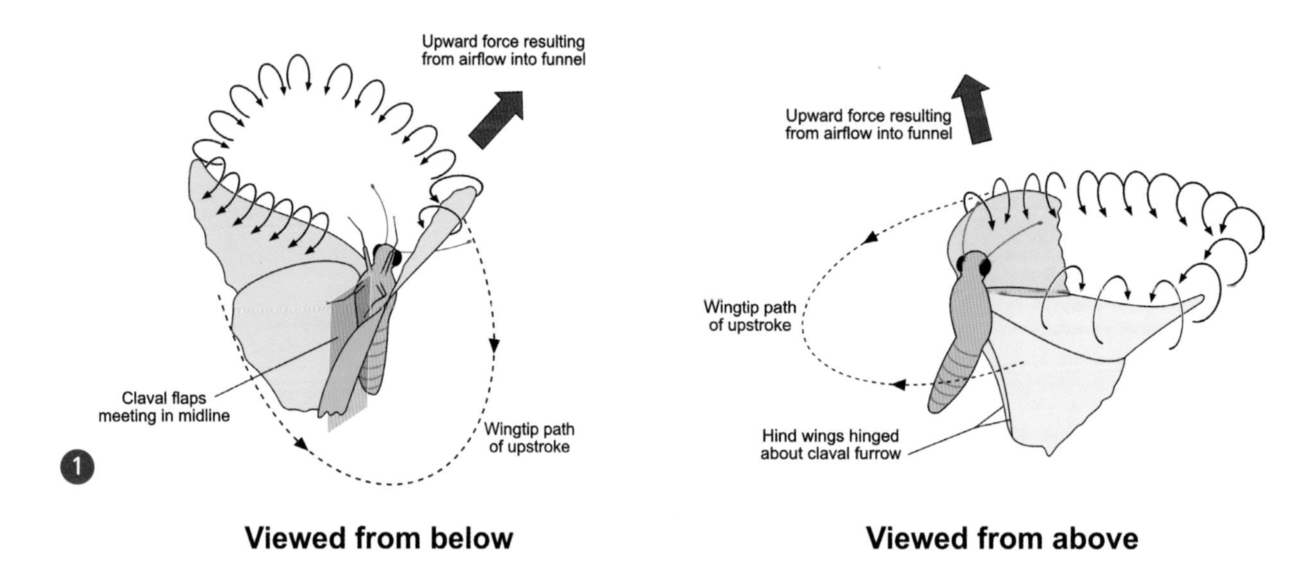

1 "Tunnel effect" from circular sweep of curved leading edge on the upstroke of a butterfly

2 The tiger butterfly (*Danaus genutia*), found in Northern India, is full of poison.

Connected wing movement

The section on birds described wings producing lift by angle and shape (see pages 52–53). This is true also of insect flight (figure 1). Butterflies and moths are called Lepidoptera and include approximately 180,000 species of these amazing creatures, which represents 10% of all the species of living organisms. Lepidoptera (and butterflies in particular) have their hindwings and forewings coupled together (figure 2) and, though not fast fliers, are much more agile than first impressions may give. They are not simply passive to the wind forces around them, but actively control their flight by generating vortices (spinning air) so that, as with hummingbirds (see page 62–63), they are able to manipulate these to aerodynamic advantage.[5]

Though their flight speed may not be more than 1 m/s in feeding flight as they go from flower to flower to collect nectar, and at most 2–3 m/s on migratory flight, there is a surprising sophistication in their wing movement. At the top of the upstroke (figure 3) there is a "clap

5 A.K. Brodsky, "Vortex Formation in the Tethered Flight of the Peacock Butterfly *Inachis Io (Lepidoptera, Nymphalidae)* and Some Aspects of Insect Flight Evolution," *Journal of Experimental Biology*, 161, 1991, 77–95.

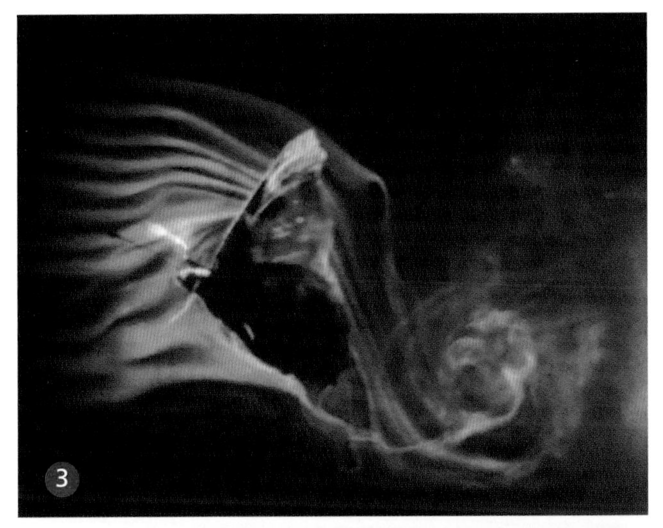

3 Tethered peacock butterfly (*Aglais Io / Inachis Io*) in a wind tunnel with smoke to make the flow visible. The wing movement is at the top of the upstroke. Notice the big drop in the flow (called "downwash") behind the wing which gives a large amount of lift. *(from Brodsky, 1991)*

4 A tiger swallowtail butterfly *(Photo by Andy McIntosh)*

and fling" movement which Lighthill[6] and Ellington[7] described in their work on insect flight. This causes a rush of air as the wings part into the vacuum which is temporarily created and this sets up a strong vortex as a result. Then the wings together are twisted downward (pronation) on the downstroke, but at the maximum part of the downstroke, as the wings now come into the bottom of the upstroke, the wings are twisted upward (supination) so that the vortices from one wingstroke cause an almost horizontal vortex ring to be made as seen in figure 1. This is the key to much of insect flight, and not just that of butterflies. Insects effectively feel and swim through a sea of vortices which they use to their advantage, much like we feel the water as we swim.

Wing scales, patterns and coloring

The wings of butterflies are so delicate that touching them will destroy the tiny scales which make up the structure of their wings. The scales, though light in weight, are extremely good aerodynamic surfaces and, as with birds, shimmering blues and greens (iridescent colors) are produced by the light being refracted and reflected through transparent and layered surfaces. Indeed, it has been shown that by artificial selective breeding, the structural color of butterfly wings can be changed from brown to violet.[8]

Some of the patterns on butterfly wings are extraordinary (figures 4 and 5) and butterflies deter predators in different ways. Some have false eyes in the patterns on their wings, making them look ominous; some are poisonous because their caterpillars feed on leaves with a high level of toxins while gradually building up their own immunity, leading to toxicity in the adults. An example of such a butterfly is the tiger butterfly (figure 2) in North India which, as a caterpillar, feeds on toxic herbs and creepers from the family *Asclepiadaceae*. Another

6 M.J. Lighthill, "On the Weis-Fogh Mechanism of Lift Generation," *Fluid Mech.* 60, 1973, 1–17.

7 C.P. Ellington, "The Aerodynamics of Hovering Insect Flight. IV. Aerodynamic Mechanisms," *Phil. Trans.*, Roy. Soc. Series B, Biological Sciences, 305 (1122), 1984, 79–113.

8 Bethany R. Wasik, Seng Fatt Liew, David A. Lilien, April J. Dinwiddie, Heeso Noh, Hui Cao, and Antónia Monteiro, "Artificial Selection for Structural Color on Butterfly Wings and Comparison with Natural Evolution," PNAS doi: 10.1073/pnas.1402770111, published online August 4, 2014. The structural color was achieved by artificially selecting for a changed thickness of the lower wing scales of butterfly wings, which then changed the way the light was reflected.

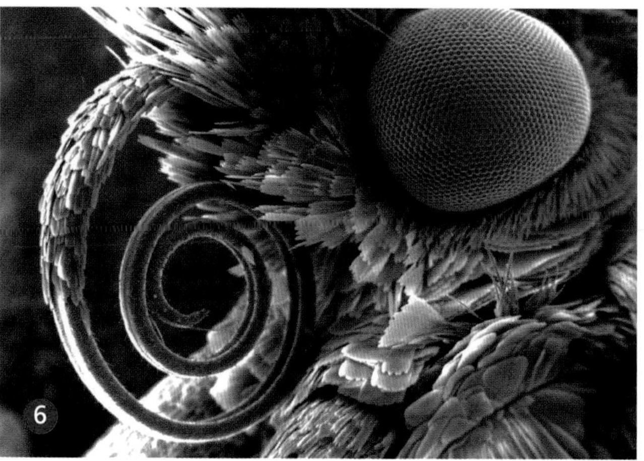

example is the cinnabar moth that feeds on ragwort and has a striped jersey effect for camouflage.

The life cycle of butterflies and moths — metamorphosis

All butterflies and moths go through a chrysalis stage in their life cycle where the caterpillar changes into something different. The creature entering the chrysalis has many legs, no wings, a mouth, a diet of leaves, and a thick body limited to crawling on vegetation. Incredibly, the new creature that emerges from the chrysalis has wings and antennae, and feeds on nectar using a special tube called a proboscis (figure 6). This process is called metamorphosis (meaning, "great change").

The butterfly mates and lays eggs to continue the species, while the caterpillar is incapable of either. Consequently, in any supposed evolutionary "story" the evolutionist must have two body plans that suddenly emerge at the point where an ancestral life form makes the step to becoming a creature with metamorphosis! This particular evolutionary transition cannot be developed slowly, as both stages in the life cycle — and both body plans — are essential in order for the creature to reproduce. This is contrary to the whole evolutionary thesis which is that any change is small.

The real observable science of the butterfly and moth denies any such slow evolution and shows what all design engineers know — that there are systems which are so complicated that all the workings are interrelated and that nothing will work unless everything works together. As referred to with dragonflies, this is irreducible complexity.

Feeding proboscis like a portable straw

Figure 7 shows the hummingbird hawk-moth and clearly visible is the proboscis which all butterflies and moths use for feeding. The proboscis has two separate halves held together by hooks, but it separates when coiled up and this also means it can be cleaned. It actually has its own muscles for coiling (figure 6) and uncoiling. Each half is an open concave trough, so that when the two halves are joined together (see figure 6) a complete tube is formed and this is used for sucking up nectar. The suction is performed by contraction and expansion of a sac inside the head.

Migration

One of the most extraordinary features of butterflies is their migratory performances. In the United States, the monarch butterfly is famous for making a journey of approximately 4,800 miles to overwintering sites in Mexico with a cycle of 4–5 generations in the journey. Some monarch butterflies even fly across the Gulf of Mexico. It is not yet understood how the next generation "knows" where to go — whether a map of some kind is carried as a genetic memory or some other technique is used.

In 2012 it was discovered that the painted lady butterfly colony (figure 8) flies through the UK from tropical Africa to the Arctic Circle in the spring

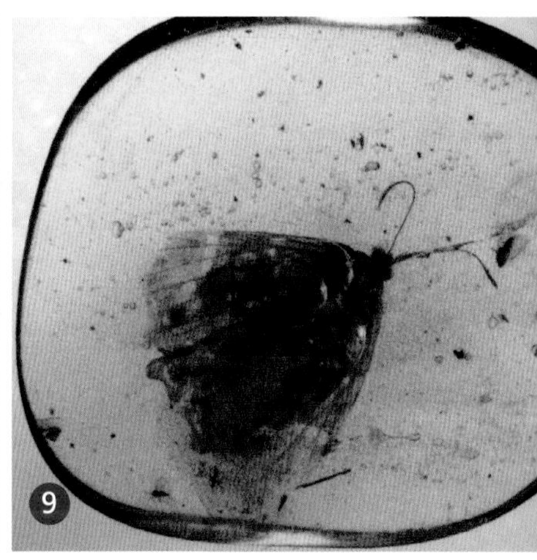

and then returns in the autumn at a high altitude of approximately 1,500 feet. The round trip is a phenomenal 9,000 miles, and involves up to six generations of these butterflies for the journey. A report was made by Butterfly Conservation at the University of York in 2012, and Richard Fox, the survey leader admitted, "The extent of the annual journey undertaken by the painted lady butterfly is astonishing. This tiny creature weighing less than a gram with a brain the size of a pinhead and no opportunity to learn from older, experienced individuals, undertakes an epic intercontinental migration in order to find plants for its caterpillars to eat."[9]

Fossils

Figure 9 shows a metalmark butterfly preserved in amber (hardened tree resin) supposedly 15 million years ago, and figure 10 is the photo of probably the best preserved butterfly fossil discovered. It was found in Colorado and is displayed in Albany State Museum NY and is supposed to be 40 million years old. Both these rare fossils actually show that Lepidoptera are essentially no different from today and to get such detailed preservation, in reality these fossils were buried fast. They are in fact silent witnesses to the truth of creation followed by a worldwide Flood, around 4,500 years ago.

9 Butterfly Conservation News Archive, October 2012. See http://butterfly-conservation.org/5183-2342/painted-lady-migration-secrets-revealed.html. Accessed March 2016.

5 The small and delicate glasswinged butterfly *(Greta Oto)* with transparent wings!

6 Scanning electron micrograph image of the proboscis of a butterfly. Notice the two halves visible in the coil.

7 Hummingbird hawk-moth with proboscis *(Macroglossum stellatarum)*

8 Painted lady butterfly (*Vanessa cardui) (Photo by Derek Ramsey*)

9 Metalmark butterfly *(Apodemia mormo)* preserved in Dominican amber and supposedly 15–20 million years old

10 A fossilized brush-footed butterfly *(Prodryas Persephone)* from Colorado supposed by evolutionists to be 40 million years old. It is also probably the same as the brush-footed butterfly we know today — so no change. But secondly, note the detail of the wings, showing that this was buried alive catastrophically and very likely is a fossil from the Flood. *(Photographed at Albany State Museum, NY, by Andy McIntosh)*

Bombardier Beetles

The extraordinary bombardier beetle emits a hot spray to ward off any would be predator — and usually wins. The spray is a mixture of caustic chemicals, hot water, and steam and is blasted out of a special nozzle which can be pointed in any direction!

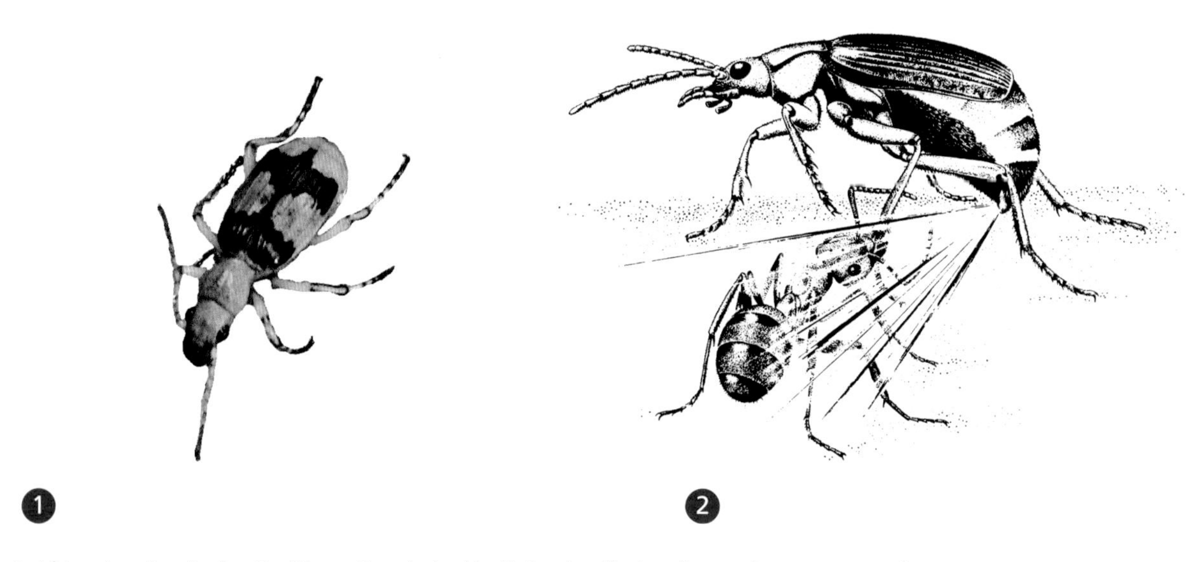

1 African bombardier beetle *(Stenaptinus insignis)* **2** Bombardier beetle spraying an ant attacker

Special defense system with moveable tank turret!

Bombardier beetles (*Carabidae Brachinini*) are found mainly in warm countries such as parts of Asia, Africa, Australia, and the United States (Florida, California). But they are also found in Europe and small colonies have even been observed in the southern part of England. They are usually not far from water and hide during the day under rocks. The bombardier beetle (figures 1, 2) ejects a mixture of chemically heated steam and noxious chemicals out of its back end through a special turret which can be moved in any direction (even twisting over its back and pointing forward — see figure 3). The whole system is used to ward off predators such as ants, birds, spiders, and frogs (figure 2). The beetle generally wins and stuns its opponent!

How does it do this? The chemicals do not come out as a continuous stream. Professor Tom Eisner in 1999 produced a seminal paper on the beetle and showed that a series of explosions is produced by the two chemicals hydroquinone and hydrogen peroxide in the presence of two catalysts: catalase and peroxidase. (A catalyst makes the reaction go much faster but does not actually undergo chemical change itself.) In a clever experiment, Eisner filmed a firing tethered African bombardier beetle, and then played it back in slow motion. Through this he showed that about 500 explosions were given off per second and that repeated blasts, each lasting 2–3 seconds, were emitted from the beetle (figure 3).

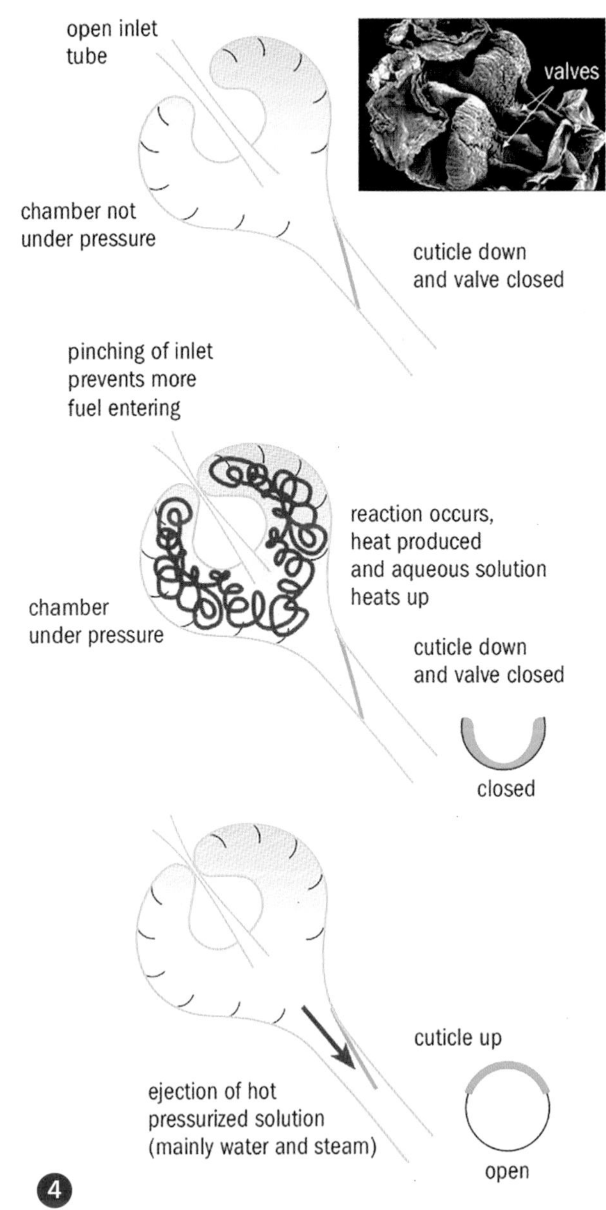

This is like firing an automatic machine gun with repeated bursts.[10]

The author (Andy McIntosh), inspired by the beetle, realized there was a clever design to be discovered, and work began at Leeds University UK in collaboration with Eisner. We showed that these blasts were controlled by a unique valve system, where not only was there an inlet valve that closed under high pressure, but that there was also an outlet valve which opens at high pressure (figure 4). As explained on page 86, this leads to a violent flash evaporation event where almost instantaneously the liquid (mostly water) expands to steam. Because a given mass of steam occupies

3 A bombardier beetle demonstrating the turret which enables it to fire over its back
(Photo reproduced under Photolibrary, London, licence 4511)

4 Bombardier beetle valve system
(Diagram reproduced from original in Physics World, 21(4), April 30, 2008)

10 D.J. Aneshansley and T. Eisner, "Spray Aiming in the Bombardier Beetle: Photographic Evidence," *Proceedings* of the National Academy of Sciences USA, 96, p. 9705–9709, 1999.

5 The V1 Flying Bomb was powered by pulse combustion, just as the bombardier beetle, and was nicknamed the "Doodlebug"

about 1,700 times the volume for the same mass of water, this ejection is with great force and carries with it much of the remaining water as well, along with the caustic chemicals. The spray has been shown to reach easily 20 times the body length of the beetle in distance.[11] (See the sequences in the film of the David Attenborough series *Life*,[12] which shows a bombardier beetle successfully warding off an ant attack).

Tiny combustion chamber

Dissections of the beetle's rear end have shown a lot more detail about its sophisticated chemical defense system. Before the two chemicals react in the tiny 1mm long combustion chamber, they travel down a very thin tube together where the catalysts either are secreted or possibly are in crystalline form. The catalysts act on the hydrogen peroxide which then converts to water/steam, thus liberating an oxygen atom for every molecule of peroxide. This then combines with hydrogen molecules released from the hydroquinone. The heat from the strong hydrogen/oxygen reaction causes the rest of the chemicals to react, and the expanding steam causes the vapor explosion.

Initial investigations of the chamber itself suggest that the chamber structure is of special heat-resistant material so that the beetle does not cook itself! The tubes both leading in and out of the 1 mm combustion chamber, as well as the chamber itself, are all totally separate to the digestive tracts of the beetle.

The valve system is a passive response system, such that the valves are operated by changes in pressure. Referring to figure 4, when the combustion chamber is empty (top diagram) and at atmospheric pressure, the inlet tube is open, allowing the reactants to enter the chamber, and the exit tube is closed by a membrane that blocks the bottom part of the tube. Once the chamber is full and the chemicals react (middle diagram) the extremities of the chamber itself, which is shaped like a boxing glove, pinch the inlet tube shut. As the chemical reaction in the chamber progresses, heat is generated and the pressure in the chamber increases until the membrane is forced open near the bottom of the exit tube (bottom diagram). The hot pressurized fluid is then ejected, the pressure in the chamber then drops, the inlet reopens allowing more reactants into the

11 N. Beheshti and A.C. McIntosh, "The Bombardier Beetle and Its Use of a Pressure Relief Valve System to Deliver a Periodic Pulsed Spray," *Bioinspiration and Biomimetics* (Inst of Physics), 2, 57–64, 2007.
A.C. McIntosh, "Combustion, Fire, and Explosion in Nature — Some biomimetic Possibilities," *Proc.* IMechE Vol. 221 Part C: J. Mechanical Engineering Science, 1157–1163, 2007.
A.C. McIntosh and N. Beheshti, "Insect Inspiration," *Physics World* (Inst of Physics), 21(4), 29–31, April 2008.

12 Martha Holmes, Rupert Barrington, David Attenborough (narrator), BBC "Life," series 6 "Insects," 2009.

6 Experimental prototype based on the bombardier beetle for spraying small droplets of fluid and vapor

chamber, and the process is repeated until all of the reactants have been exhausted.

This process is called "pulse combustion" and is used by some engines to give thrust. The most infamous example of this was the V1 "Doodlebug" Flying Bomb (figure 5) of World War II which was used to great effect in 1944 against London and the southern counties by Hitler. In the case of the V1, the fuel was petrol burning in air. At that time, few appreciated that a similar combustion system was already in use by the bombardier beetle, not for propulsion, but for spraying its attackers.

Bio-inspiration from the bombardier beetle

Research which began at the University of Leeds has continued to develop a spray system (figure 6) which is based on the technique used by the beetle. The testimony of this author (Andy McIntosh) is that, contrary to the allegation that a belief in creation closes down research, it was precisely because I knew the beetle chamber was *designed* that led me to these investigations.

It was clear there were design features to be understood and this has led to a patented spray facility which heats water in a special chamber (approximately 20 times the size of the bombardier beetle chamber) where inlet and outlet valves are controlled electronically to open and close at an assigned time. This is an active control system using no chemistry in contrast to the passive system of the beetle which uses chemical heating. However, the valve system itself is very similar to that used by the beetle, and one of the prototypes is displayed in figure 6. In 2010, our work won the Times Higher Education award for the most outstanding contribution to innovation and technology. Possible uses for this are for developing spray systems for fuel injectors in car and truck engines, for fire extinguishers, and for fragrancers used in meeting rooms.

Design features of the bombardier beetle

Any system involving combustion has to be very carefully designed because combustion is dangerous! And it is clearly an example of irreducible complexity since the combustion system will not work unless all the design features are in place. Some of the unanswered questions arising from the bombardier beetle research are: In what form are the catalysts? How does the beetle sense the direction of attack? How does the moveable turret work that directs the exhaust? How are the chemicals hydrogen peroxide and hydroquinone produced?

The very fact that many of these questions are still not answered shows the irreducible complexity of this creature and its evident design!

Ants

Ants are some of the most prolific insects on the earth. They are very strong for their size, hugely industrious, and extremely robust in their ability to keep their colonies alive against any predators.

1 Leafcutter ants *(Atta colombica)*

Ants — social creatures in all sizes

It is estimated that there are 22,000 species of ants and that little more than 12,500 have been classified. The technical term for the ant family is Formicidae, belonging to the order Hymenoptera along with wasps and bees. Perhaps one of the most staggering facts is that the ants account for 15–20% of the total terrestial animal mass and actually exceeds that of vertebrates.[13] The largest ants are the carpenter ants which can measure up to 0.5 inch in length with their queens being 0.8 inches long. Ants live everywhere except Antarctica, some remote islands, Greenland, and Iceland. But where they do reside, their numbers are overwhelming. National Geographic states, "One hectare (10,000 square meters or approx 2.5 acres) of land in the Amazon rain forest can contain eight million ants or more. A study in the savanna of Côte d'Ivoire showed that a hectare there harbored 20 million."[14]

Ants dwell in colonies and have a clear social structure. A mature colony of carpenter ants usually contains around 3,000 adults,[15] but some species have been known to contain up to 100,000. The colonies are composed of different "castes." A typical parent colony contains a queen, the queen's brood, and workers, both minor and major. The size of worker ants determines their responsibilities. Minor workers are the smallest members of the

13 T.R. Schultz, "In Search of Ant Ancestors," *Proceedings of the National Academy of Sciences* 97 (2000) (26): 14028–14029.

14 Alice S. Jones, "Fantastic Ants — Did You Know?" *National Geographic Magazine*. Archived from the original on July 30, 2008, http://ngm.nationalgeographic.com/2007/08/ants/did-you-know-learn. Accessed March 2016.

15 http://www.orkin.com/ants/carpenter-ant/carpenter-ant-size/. Accessed March 2016.

2 Army ants *(Eciton vagans)* on a raid carrying home larvae from a wasp's nest

colony, and their tasks are to take care of the young and forage for food. Major workers are larger and serve as soldiers to defend against predators.

The life cycle of ants

Ants start their lives from eggs. If the egg is fertilized, the progeny will be female; if not, they will be male. Ants develop through a larva stage, then a pupal stage, before emerging as an adult.[16] The larva is essentially immobile and is fed and cared for by workers who give food by regurgitating liquid nutriment held in their crops. The larvae then grow through a series of four or five molts and enter the pupal stage. The pupa will become either a queen, or a worker (both female). The type of ant and the different castes of workers is determined by the nutrition the larvae obtain, though the determination of caste continues to be a subject of research.[17] Winged male ants, called drones, emerge from unfertilized pupae along with the usually winged breeding females from fertilized pupae. Some species, such as army ants, have wingless queens.

A new worker spends the first few days of its adult life caring for the queen and young. She then graduates to digging and other nest work, and later to defending the nest and foraging. Ant colonies can be long-lived. The queens can live for up to 30 years, and workers live from one to three years. Males, however, survive for only a few weeks.

Leafcutter ants

These ants cut the leaf with their mandibles (jaws) and then carry pieces of leaf, far larger than their own size, to where the colony maintains a fungus used for food (see figure 1). The central mound of their underground nests can grow to more than 98 feet across, with smaller, radiating mounds extending out to a radius of 260 feet, taking up 320 to 6,460 square feet, and containing eight million individuals!

Army ants

There are actually several different species of Army ants (sometimes called soldier ants — figure 2).[18] Army ants live in temporary nests and seldom make underground burrows like other ants. They live in Africa, South America, and Central America. Their

16 https://en.wikipedia.org/wiki/Ant. Accessed March 2016.

17 Kirk E. Anderson, Timothy A. Linksvayer, and Chris R. Smith, "The Causes and Consequences of Genetic Caste Determination in Ants (Hymenoptera: Formicidae)." Myrmecol. News 11: 119–132, 2008. See http://myrmecologicalnews.org/cms/index.php?option=com_content&view=category&id=250&layout=default. Accessed March 2016.

18 http://www.insects.org/entophiles/hymenoptera/army-ants.html. Accessed March 2016.

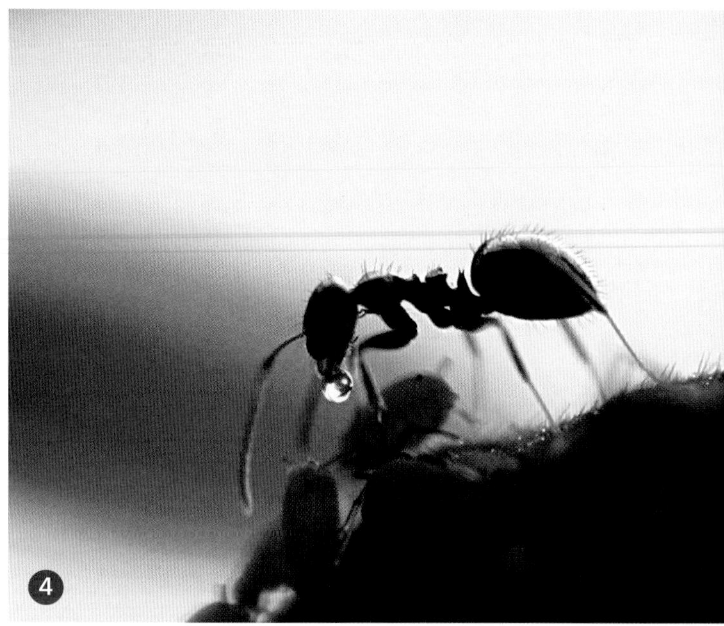

temporary nests, or bivouacs, are places where the ants rest between their hunting raids. The bivouac is typically a hollow log, or the colony might hang from a tree limb. Thousands of workers will connect their legs and their mandibles and make an enclosed hammock for the queen ant, and sometimes the workers enclose the immature ants inside the hammock as well. The army ants stay in the bivouac for a few weeks, and only when the queen comes out of the resting place does the colony then start to move.

Some species of army ants move in a line, while others migrate in a fan-shaped wave. When this happens, many thousands of ants all move at once and eat everything in their path. The soldier ants march at the side of the column to defend the queen, and some of the workers carry the immature ants. Other workers gather all the food that they can find. As they move, the workers kill every insect, spider, snake, and lizard in their path. Birds and animals hear the ants marching and flee! As they march, the ants can climb trees or shrubs, and have even been known to go through houses during the march with residents scrambling to safety and removing poultry and livestock to a safe location. The ants eat everything that does not run or fly away. Some of these army ant species raid other ant colonies and take their larvae as food for their own colonies.

Communication

Ants communicate by using chemicals called pheromones, but also by sounds and touch. They perceive smells with their long, thin, and mobile twin antennae. These provide information about the direction and intensity of scents. Since most ants live on the ground, they use the soil surface to leave pheromone trails that may be followed by other ants. That is why when you see an ant trail it is always following the route placed by the scout who went ahead. If the scent is removed by water or some other chemical that neutralizes it, the ants become confused.

The Bible and the ants

Ants are incredibly industrious and social creatures (honeypot ant, figure 3; farming aphids, figure 4; stitching leaves. figure 5). The superb example of work and organization of ants is referred to in the Bible: "Go to the ant, you sluggard! Consider her ways and be wise, which, having no captain, overseer or ruler, provides her supplies in the

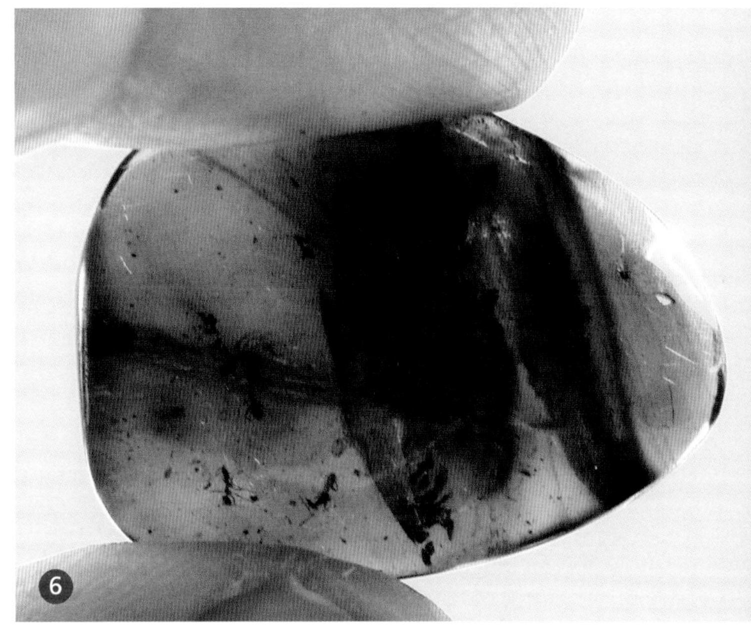

summer, and gathers her food in the harvest" (Proverbs. 6: 6–8).

Ants were created to show the combined social and organizational structure that maintains vast numbers of creatures in an ordered colony where each ant knows what it is meant to be doing for the welfare of the whole group. To suggest that these creatures have evolved is not borne out by the evidence of the sophisticated behavior of present colonies, where the whole colony is dependent on each ant doing its assigned work and taking care of each pupa. This is particularly evident in the work of leafcutter ants and weaver ants. The highly ordered structure of ant societies bears testimony to the God who made them.

Neither is there any evidence of ant evolution in the fossil record. Figure 6 shows amber (hardened tree resin) encasing trapped ants supposedly from 90 million years ago; they are obviously no different from modern ants. The ant kind has always been an ant and has not evolved from anything else. Some ants may have changed their behavior and some speciation will have occurred, but the basic family Formicidae has always been ants. This particular example in figure 6 is likely to have been formed in the devastation of the biblical Flood.

The Bible also uses the ant to remind us that we should plan ahead: "The ants are a people not strong, yet they prepare their food in the summer" (Proverbs 30:25). It is a foolish person who does not observe the ant and learn that our season also comes to an end, and we need to prepare for eternity.

3 Honeypot ants *(Myrmecocystus mexicanus)* hanging on the ridges of their colony where their bloated abdomens become larders of sugar-rich nutrient in times of food scarcity for this species living in hot deserts such as Mexico. *(Photo by Greg Hume at Cincinnati Zoo)*

4 Ant receiving honeydew from an aphid. Aphids are farmed by some ants who tap them with their antennae to have them secrete the honeydew that they then feed on.

5 Weaver ants stitching leaves together in highly coordinated team work. The leaves are sewn together using silk from the heads of maturing larvae.

6 Ants fossilized in amber supposed to be 90 million years old. The ants are essentially no different from modern ants.

Bees

Bees are amazingly industrious and have highly organized temperature-controlled hives with carefully constructed homes for raising the larvae.

1 The honeybee *(Apis mellifera)* flying into the hive carrying pollen in its basket **2** European honeybee extracting nectar

The honey bee

There are about 20,000 known species of bees in seven to nine recognized families. Bees, like ants, are social creatures and have some of the characteristics of ants. They are both classified as part of the order Hymenoptera. However, this grouping of two creatures together under one classification is based on an evolutionary worldview and does not mean that there is actually any common descent of these creatures.The honeybee worker has many distinctive features. The way it flies (figure 1), gathers its nectar (figure 2) to make honey, produces beeswax from special glands, and the pollen balls it carries for the growing larvae to feed on, all show the wonder of a specially designed creature. Here we discuss mainly the honeybee and later the bumblebee (bombus). As with ants, there is a queen of the colony, and after mating the queen bee lays eggs which, when fertilized, are female (workers) and if unfertilized, are male (drones). The queen stores the sperm and determines which sex is required at the time of egg laying.

The hive

Though hives for bee farming are made artificially so that bee keepers can control the honey production, in the wild the bees will make a beehive naturally (figure 3) of interlocking hexagons. This was proven by Hales to be the most efficient way of using the beeswax to get as many larvae cells in an area as possible.[19]

In 1999, Keith Devlin reported, "Not until the advent of close-up film techniques did scientists know for certain how bees build their honey stores. It is a remarkable feat of high precision engineering. Young worker bees excrete slivers of warm wax, each about the size of a pinhead. Other workers take the freshly produced slivers and carefully position them to form vertical, six-sided, cylindrical chambers (or cells). Each wax partition is less than 0.1mm thick, accurate to a tolerance of 0.002mm. Each of the six walls is exactly the same width, and the walls meet at an angle of precisely 120 degrees,

19 Thomas C. Hales, "The Honeycomb Conjecture," Discrete and Computational Geometry, 2001, 25 (1): 1–22.

producing one of the 'perfect figures' of geometry, a regular hexagon."[20]

The bees somehow "know" that the best packing geometry is not squares but hexagons!

The hives are also temperature controlled so that if too hot, worker bees fan the air with their wings to cool the hive. Perhaps one of the most stunning discoveries of all is that the distance and direction of the best food supply is communicated to the other worker bees by a "waggle dance" (figure 4), which Austrian scientist Karl von Frisch discovered in 1944 but was not recognized until the 1950s and 1960s.[21] The angle to the vertical indicates the direction angle relative to the sun, and the number of waggles

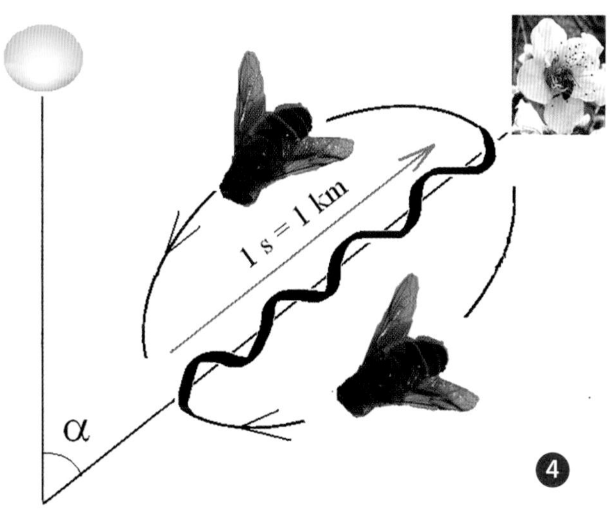

3 Natural beehive and honeycomb.
(Photo by Greg Hume at Cincinnati Zoo)

4 Honeybee workers can navigate, indicating the range and direction of food to other workers with a waggle dance.

20 Keith Devlin, "Secrets of the Beehive," *The Guardian*, August 26, 1999, https://www.theguardian.com/science/1999/aug/26/technology. Accessed March 2016.

21 Karl von Frisch, *The Dancing Bees* (Harcourt, Brace & World, 1953), p. 93–96. Also see K. von Frisch, *The Dance Language and Orientation of Bees* (Harvard University Press, 1967).

5 Bumblebee *(Bombus hortorum)* carrying its precious cargo of pollen

indicates the distance to fly. Von Frisch and later authors demonstrated that bees can recognize a desired compass direction in three different ways: by the sun, by the polarization pattern of the blue sky, and by the Earth's magnetic field.[22]

The staggering complexity of honeybee behavior shows the mind of an engineering genius behind such skills which involve sophisticated communication and mapping systems. Truly, we should give God the glory and exclaim "O Lord, how great are Your works! Your thoughts are very deep!" (Psalm 92:5).[23]

22 Randolf Menzel et al, "Honeybees Navigate According to a Map-Like Spatial Memory," PNAS, 2005, 102 (8): 3040–3045.

23 Just as Samuel Morse, in another brilliant feat of communication, in the first dispatch of a long distance telegraph from Washington to Baltimore and back, on May 24, 1844, telegraphed "What hath God wrought!" (Numbers 23:23 KJV).

The bumblebee

One of the favorites in the bee world is the bumblebee (*Bombus*). Bumblebees (figures 5, 6) form colonies of between 50 and 400 individuals; these are small compared to honeybee hives which hold about 50,000 bees. As with all bees, bumblebees beat their wings at about 200 cycles per second. Their thorax muscles do not contract on each nerve impulse, but rather they vibrate like a plucked rubber band at a natural (resonant) frequency. This is efficient, since it lowers the energy consumption, as the insect's motor nerves (operating the system of muscles and wings) could not fire actively at 200 hz (cycles per second).[24] All bees fly remarkably efficiently because of the

24 C.W. Scherer, "Fastest Wing Beat," *Book of Insect Records* (University of Florida), http://entnemdept.ifas.ufl.edu/walker/ufbir/chapters/chapter_09.shtml. Retrieved October 29, 2015.

6 Bumblebee — *Bombus terrestris*

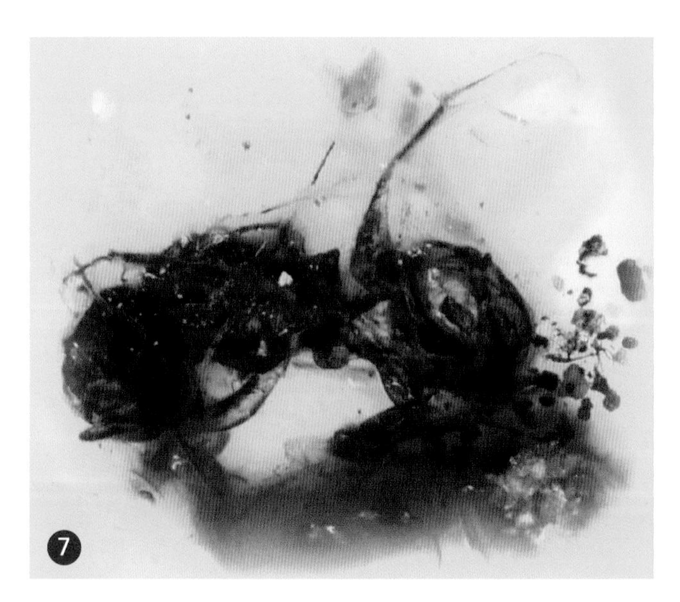

7 Bee *Melittosphex Burmensis* trapped in Lower Cretaceous amber, supposedly 80–90 million years old. However, there is no evidence of any major change; this was catastrophically trapped, probably in the Flood.

dynamic twisting of their wings and manipulation of vortices. The motion is similar to that described for hummingbirds and dragonflies (see pages 62–63, 76).

Figure 7 shows that amber fossils confirm that there has been no bee evolution. They have always been bees!

It has been suggested that because of its relatively small wings and large body, the bumblebee should not be able to fly, but there is a clever trick that all insects use which is the twisting of the wings as they go through a beating cycle. We discussed this earlier (see pages 80–81) with butterflies and moths and showed that supination (twisting upward on the upstroke), and pronation (twisting downward on the downstroke) give dramatically more lift, which is why the bumblebee manages perfectly well, thank you!

Dung Beetles

Dung beetles are extraordinary. We would be overrun with dung from many animals were it not for these humble creatures who make it their function in life to roll muck and break it up!

The beetle that loves muck

Beetles actually make up about one-third of all known insect species. Scarab beetles, commonly known as dung beetles, have the Latin family name of Scarabaeinae and feed exclusively on faeces (dung), and comprise about 5,000 species.[25] Many dung beetles (figure 1), known as "rollers," roll dung into round balls, which are used as a food source for their young in brooding chambers. Others, known as "tunnelers," bury the dung wherever they find it. A third group, the "dwellers," neither roll nor burrow — they simply live in the manure! These beetles can grow to 1.1 inch long and 0.8 inch wide and play a remarkable role in agriculture. This is because by burying and consuming dung, they improve the recycling of nutrients in the soil and they also protect livestock, such as cattle, by removing the dung which, if left, can breed pests like flies. So these beetles are very important and function well in dealing with the mess that otherwise would be left by the larger animals. Dung beetles can eat more than their own weight in 24 hours.[26]

25 https://en.wikipedia.org/wiki/Dung_beetle. Accessed March 2016.

26 http://www.britannica.com/animal/dung-beetle. Accessed June 2016.

1 Dung beetle *(Scarabaeus viettei)*

2 Two dung beetles *(Scarabaeus laticollis)* fighting over a ball of dung.

3 Dung beetle with two balls of dung

Battles over dung

The dung beetle rolls the dung, always following a straight line despite all obstacles, back to its home where the larvae are hatching and feeding on the dung. For most dung beetles it is just the male who does the pushing, and the female sometimes hitches a ride on top of the dung! He uses his back legs to push the ball; occasionally male and female beetles do this together. Dung beetles are so keen to get the dung that battles can emerge (figure 2) between them for ownership. They even occasionally try to steal the dung ball from another beetle, so they have to move rapidly away from a dung pile once they have rolled their ball, to prevent it from being stolen. They can roll up to ten times their weight. Such is the determination to get at the feces balls, it has been known for dung beetles to roll two at once (see figure 3).

Navigation by the stars!

In order to orient themselves, before rolling the dung ball the beetles will often get on top of the ball and perform a "dance" during which they locate light sources to use for orientation; they then work out which direction they need to roll. Remarkably, it has been confirmed that this humble beetle actually maps its direction by the stars (figure 4) of the Milky Way! Marcus Byrne of the University of the Witwatersrand in South Africa showed this by using a planetarium.[27] It is believed that a bright moon can also be used by the beetle.[28] It is the only insect known to orientate itself by the galaxy.

We are reminded of Job 12:7–8, "But now ask the beasts, and they will teach you . . . or speak to the earth, and it will teach you." The beetle may be a lowly creature, but the sophistication of the dung beetle is a lesson to our proud hearts. Even these creatures are fearfully and wonderfully made.

27 Marie Dacke, Emily Baird, Marcus Byrne, Clarke H. Scholtz, Eric J. Warrant, "Dung Beetles Use the Milky Way for Orientation," *Current Biology*, 2013.

28 Wits University, "Dung Beetles Follow the Milky Way: Insects Found to Use Stars for Orientation," *Science Daily*, January 24, 2013. https://www.sciencedaily.com/releases/2013/01/130124123203.htm. Accessed March 2016.

4 Dung beetles have been shown to be able to orientate themselves by using the Milky Way galaxy. No other creature has been shown to do this.

5 The dung beetle as a symbol of the gods in Egyptian religions. *(Courtesy of the British Museum)*

Dung beetle and the Bible

For many centuries across the Ancient Near East the sacred scarab (*scarabaeus sacer*) was seen as a symbol of new life as the beetle grubs and vegetation came from the ball of dung. It was associated with the creator god Atum. It also represented the sun god Khepri being reborn each morning as the sun rose from the Earth. The scarab appears as large monuments (figure 5), tiny amulets and figures on inscriptions and tombs; it was especially popular in Egypt. Significantly, however, it is not mentioned once in the Bible, even though some try to suggest Israel copied its religion from the surrounding nations.

06 Stars and Planets — The Sun

The sun is a stable star and is at just the right surface temperature to emit light and heat to sustain life on Earth.

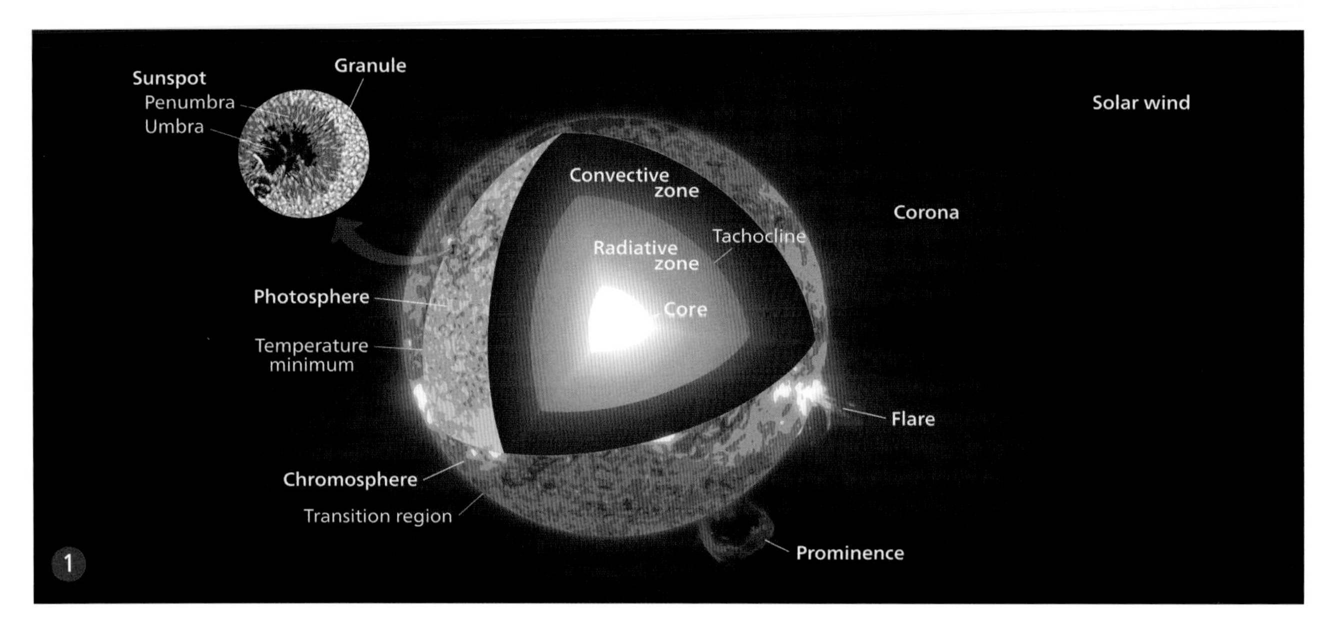

1 The structure of the sun. All features drawn to scale. *(Original drawing by Kelvinsong)*

The sun is essentially made from mostly hydrogen and helium atoms but without their electrons (termed a plasma).[1] Its enormous energy supply is primarily produced by fusion of hydrogen into helium nuclei (alpha particles) in its inner core.[2] The sun's diameter is 865,000 miles, which is 110 times the size of Earth (diameter 7,890 miles) and is 93 million miles distant from us.

The inner core (see figure 1), with a temperature above 15 million degrees Celsius, extends to approximately a quarter of the radius of the sun, i.e., approximately 108,000 miles from the sun's center. The power of the sun is a stable 3.846×10^{26} watts (a watt is a measure of power in energy units [joules] per second), and unlike other stars it fluctuates very little. To understand how large this is, the world population consumes about 12 terawatts (12×10^{12} watts), so the sun produces 3 million million times the total power that we currently use on Earth!

Beyond its core, extending to about 2/3 of the radius of the sun, is the solar radiative zone (figure 1). Here, though still millions of degrees, the temperature is not high enough for fusion. Then the outer 1/3 of the radius of the sun is the convective zone with ionized gas (plasma) at a lower temperature of the order of thousands of degrees. Observations show that the magnetic field of the

1 Much of this material has come from a series of articles by Jason Lisle in Institute for Creation Research *Acts and Facts*, July 2013–May 2014. *Acts and Facts* is available free of charge at the ICR web site www.icr.org.

2 Each fusion process converts approx. 0.7% of the mass of the protons to energy which heats the core. With vast amounts of energy produced each second, very small particles are also produced called neutrinos which pass out into space and pass even through dense matter. These were only finally detected in 2014 through a device deep underground in Italy which then proved that it is fusion which primarily produces the sun's power. See "Borexino Collaboration, Neutrinos from the Primary Proton–Proton Fusion Process in the Sun," *Nature* 512, 383–386, August 2014.

sun is produced here by the churning of large cells of electrically charged hot gas.

Solar eclipse

Only at a total eclipse can one see the chromosphere and the corona (see figure 2). This rare event occurs about 2 or 3 times a year in different parts of the world. A remarkable design feature is that the moon is approximately 400 times smaller than the sun but the sun is about 400 times farther from us than the moon, so we can have the situation where the sun is blocked almost exactly by the moon since its apparent (angular) size in the sky is almost the same. Partial eclipses occur over a wide band, with a much narrower band experiencing a total solar eclipse. The last total eclipse in the UK was in southwest England on August 11, 1999 (see figure 2). The next one will be on September 23, 2090!

Sunspots

The convection zone reaches the sun's surface as the photosphere, with cells of twisting gas called "granules" and a temperature of about 5,527°C/9,980°F. Locally, the magnetic field can be such that the convection is inhibited, which causes local cooling and dark spots to occur (in comparison to the rest of the sun's surface). These "sunspots" (see figure 3) grow in a cycle of approximately 11 years. As the sunspots grow, the solar power reaching Earth (irradiance) grows by 2–3 watts m^{-2}. The sunspots appear to be linked to two magnetic fields oscillating with almost, but not exactly, the same frequency.[3]

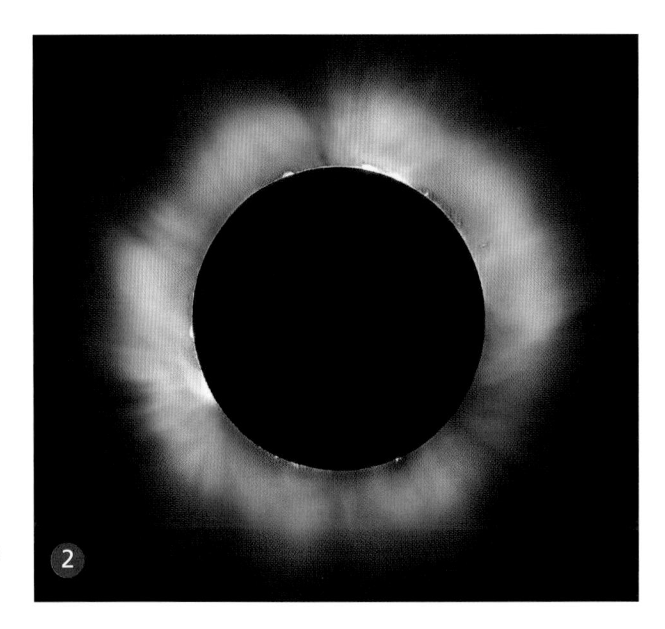

2 A total eclipse of the sun on August 11, 1999, in southwest UK. The colors that are visible in the chromosphere are very impressive and add to the strange and unusual experience seen by observers. Beyond this is the corona, a large disk of ionized gas which is actually much hotter and more energetic than the sun's surface (photosphere) and chromosphere.

Solar flares

"Solar flares" are often connected to sunspot activity. They shoot an eruption of solar material through the sun's atmosphere called a "prominence" (see figure 4). These can affect Earth's magnetic field, bringing intense geomagnetic storms of charged particles into our atmosphere, and can degrade communication signals with unexpected electrical surges in power grids. They also enhance the auroras seen in the polar regions. Nevertheless, solar flares are small compared to those of other stars of similar size which are not stable. Some stars have super flares extending far into space and such would be catastrophic for Earth. Our sun is exactly right for life.

The design of the sun

The sun shows remarkable design. Its surface temperature of 5,527°C/9,980°F is exactly right for radiating most of its energy in the visible spectrum (figure 5). If the sun was slightly cooler, there would be less intense light and the peak of the power felt

3 S. Shepherd, S.I. Zharkov, and V.V. Zharkova, "Prediction of Solar Activity from Solar Background Magnetic Field," *The Astrophysical Journal*, 795:46 (8pp), November 2014. Also see "Solar Activity Predicted to Fall 60% in 2030s, to 'Mini Ice Age' Levels: Sun Driven by Double Dynamo," *Science Daily*, July 9, 2015. The evidence is of two magnetic dynamos, one deep in the convection zone and another near the surface of the sun. The oscillation periods are near 11 years but they are not exactly the same. One is slightly longer than the other. Hence there are times of much greater activity when they act in phase with a slightly greater heating, and then the opposite when the two frequencies are out of phase which correspond to cooling such as the Maunder minimum of 1640–1700. So there is some subtle variation in the heat of the sun over decades.

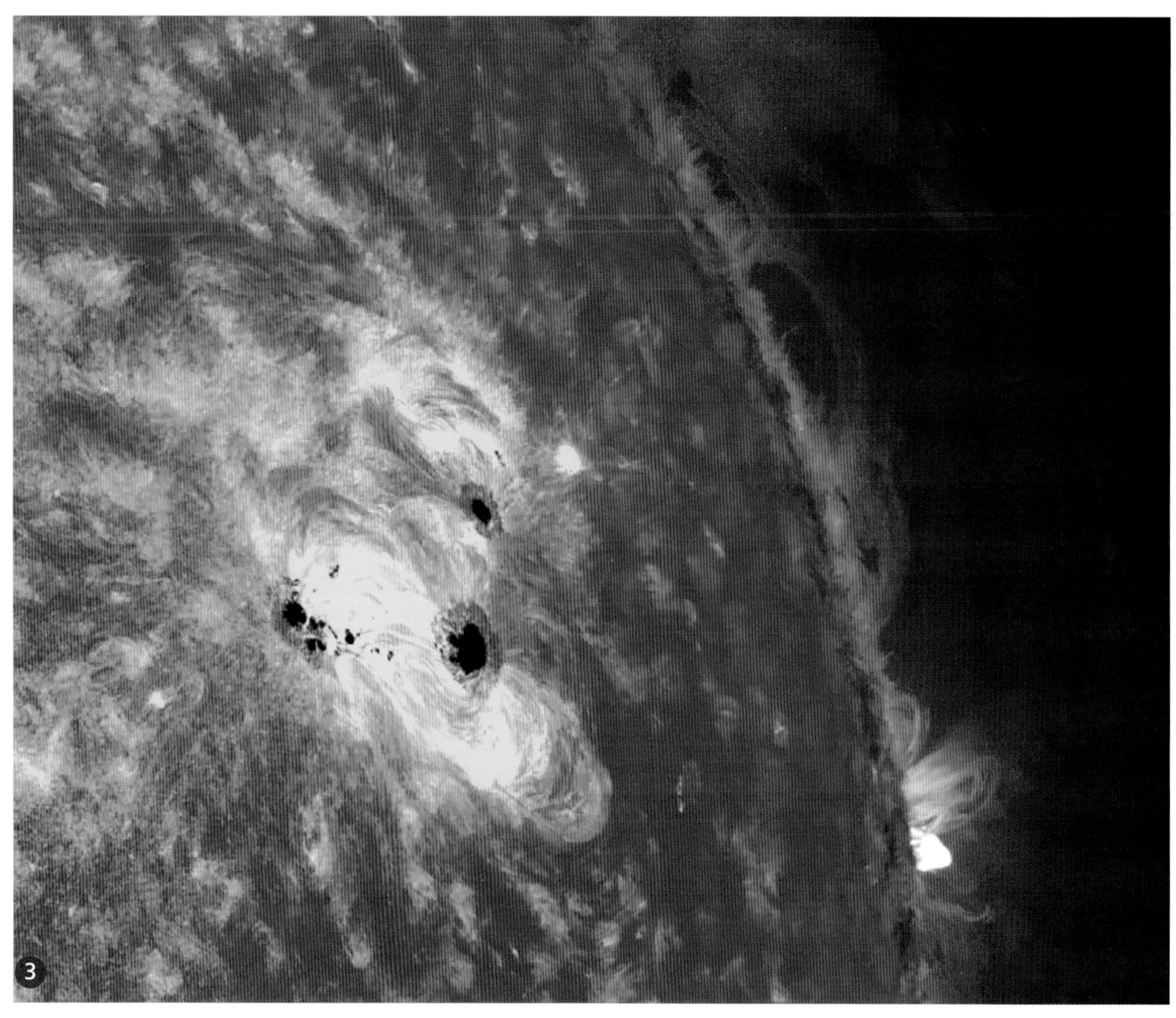

3 Fast-growing sunspots, February 2013. The sunspots themselves are many times the size of Earth. *(Photo by NASA Solar Dynamics Observatory, SDO])*

would be in the infrared region, and life as we know it could not be maintained. If the sun was slightly hotter, then the peak would be in the ultraviolet region which would be very harmful for life. Earth is also in exactly the right position to receive the right amount of heat to sustain life. Closer in to the sun, and the seas would evaporate — farther out, and the seas would freeze.

The sun's energy provided by fusion shows that the sun is not billions of years old. The evolutionary view of the sun's formation is that the sun can only have been undergoing fusion reactions for the last 5 billion years and before that it was a much fainter star. However, if that were true, then the sun would have had no power to support life (termed the faint young sun paradox[4]). For the biblical timescale with a 6,000-year-old sun this is not an issue.

Everything points to the sun being made precisely, and positioned exactly, for Earth to sustain life. The true science of the solar system agrees with Genesis.

4 J. Lisle, "The Solar System: The Sun," Institute for Creation Research *Acts & Facts*. 42 (7): 10–12, July 2013. For discussion of the faint young sun paradox, see note 4 in that article.

4 Solar giant prominence eruption 2012. *(Photo by NASA/SDO/AIA/Goddard Space Flight Center)*
5 Blackbody radiation against wavelength for different temperatures

The evolutionary hypothesis also proposes that the sun was formed by a big cloud of nebulous hydrogen and helium which collapsed under the force of gravity. However, as Jason Lisle states, "Astronomers have discovered thousands of nebulae, but no one has ever seen a nebula collapse in on itself to form a star. The outward force of gas pressure in a typical nebula far exceeds the meager inward pull of gravity. As far as we know, nebulae only expand and never contract to form stars. Even if gravity could somehow overcome gas pressure, magnetic fields and angular momentum would tend to resist any further collapse, preventing the sun from forming at all."[4]

Everything demonstrates the truth of Day 4 of the creation week recorded in Genesis 1:16: "God made . . . the greater light to rule the day. . . ."

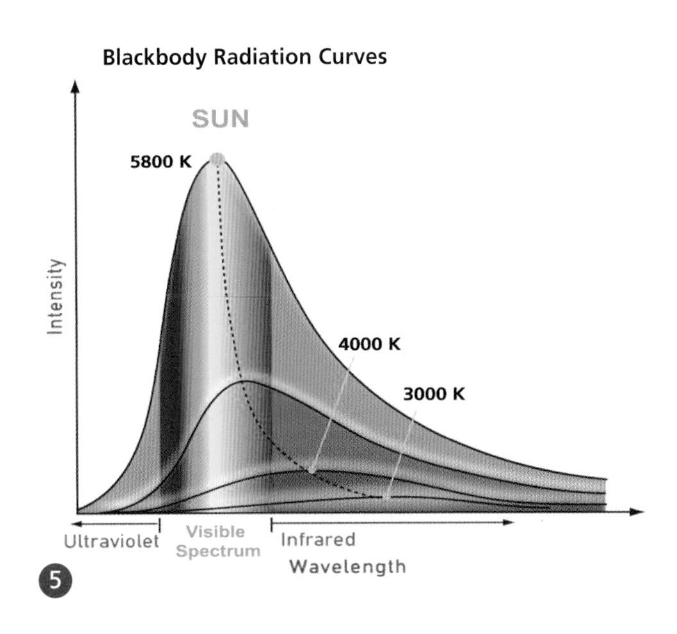

The Moon

The moon is no random satellite of Earth. Its size, position, and motion are exactly right for maintaining tides and currents. This contributes to the sustaining of life on Earth.

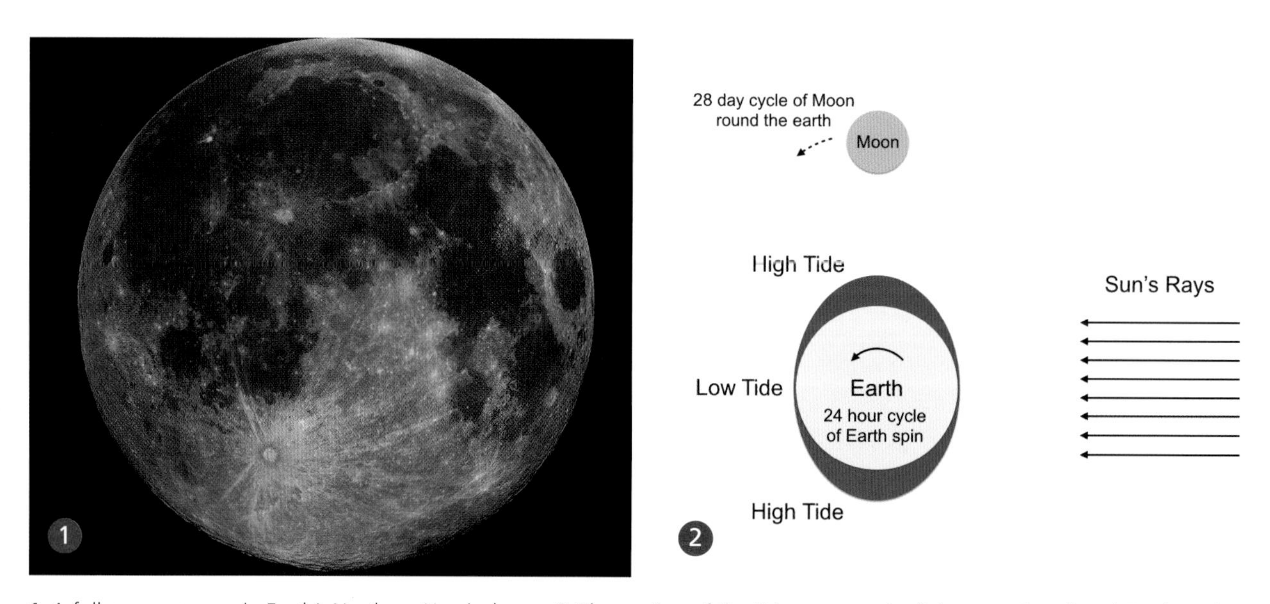

1 A full moon as seen in Earth's Northern Hemisphere **2** The motion of the tides as a result of the moon's pull and Earth's spin

The full moon is a glorious sight (figure 1). Our moon is one of the largest satellites in the solar system. Jupiter's Ganymede, Castillo, and Io, as well as Saturn's Titan, are larger, but our moon is much closer to Earth than the moons of other planets. Relative to the size of their host planet, the moon is by far the largest moon, although in terms of the angular appearance in the sky from the host planet, Jupiter's Io and Neptune's Triton come close to the appearance of the moon from Earth. The moon has approximately one-quarter of Earth's diameter and therefore approximately one-sixteenth of the area of Earth.

The moon pulls on Earth and in particular on the oceans on Earth so, as the moon moves relative to Earth (due to the spin of Earth and due to the slower movement of the moon eastward through the month), it drags the water with it. It is exactly positioned to give just the right movement of tides: too close and the tides would swamp the land; too far away and there would not be enough effect to keep the oceans moving, and the life in the sea refreshed. This is vital. The tides work by the moon pulling the sea in its direction and causing a bulge (high tide) toward and away from the moon (see figure 2). Halfway between are low tides. As Earth spins, this effect washes around Earth causing approximately two tide changes per day, and as the moon slowly moves around Earth every 28 days, the time and location of the tidal maximum changes. A further effect is that as the month progresses there will be neap tides when the sun is on the opposite side tending to cancel the moon's effect, and spring tides when the sun is adding to the moon's effect. Figure 2 shows the part of the month when the sun's position is between these points. As the year progresses, where these occur will change on the surface of Earth.

3 Buzz Aldrin on the surface of the moon when he and Neil Armstrong were there July 21, 1969

4 Lunar eclipse on September 27, 2015. This is an example of what is called a blood moon. *(Photo by Alfredo Garcia Jr.).*

Origin of the moon

Though men have walked on the moon (figure 3), understood the effects at an eclipse (figure 4), explored it (figure 5), and viewed Earth from it (figure 6), evolutionary explanations elude those who believe the moon evolved and deny creation. They propose the following arguments, none of which are workable scientifically.[5]

1. Moon fission (break away) from Earth

This suggests the moon formed out of Earth when Earth was molten in its early stages (of the order of 4.5 billions of years ago), and when Earth was cooling and rotating very rapidly. However, with such a theory, a rotation period of approximately 2.6 hours would be necessary for the rotational instability to occur. Yet a rotation period of as little as 5 hours with the moon only a few radii away would cause tides kilometers high, melt Earth's mantle, and vaporize the oceans. Also, this proto-moon would need to go through a critical distance called the Roche limit (the minimum distance at which a satellite can withstand the gravitational forces exerted on it without disintegrating). For Earth, this is approximately 11,500 miles. The moon is 21 times this distance from the Earth (238,856 miles). Therefore, it could not reach this distance without breaking up.[6] There are also important chemical differences in the geology of the moon to that of Earth (for example the moon has a low iron content), which imply they are not from the same source.

2. Capture of the moon by Earth

This view is the notion that the moon was an object caught by Earth's gravity. It is an impossible scenario, since in order for the moon to be captured intact, its speed could not exceed 131 ft/s. If its speed exceeds 8,202 ft/s, it would be put into an orbit around the sun, and in between these speeds it would be broken into particles like Saturn's rings.[7] As Jonathan Sarfati has stated, "So the moon was wandering through the

5 J.C. Whitcomb and D.B. DeYoung, *The Moon — Its Creation, Form and Significance* (Winona IN: BMH Books, 1978); see chapter 2, "Naturalistic Theories of Moon's Origin."

6 P. Goldreich, "Tides and the Earth–Moon System," *Scientific American*, 226:4, 51, April 1972; also W.H. Munk, "Once Again — Tidal Friction," *Quarterly Journal of the Royal Astronomical Society*, 9, 352–375, 1968.

7 A.L. Hammond, "Exploring the Solar System (III): Whence the moon?" *Science* 186 (4167), 911–913, 1974.

5 NASA Apollo 17 Lunar Roving Vehicle

solar system, and just so happened to be captured by Earth's gravity... but the chance of two bodies passing close enough is minute; the moon would be more likely to have been 'slingshotted,' like artificial satellites, than captured. Finally, even a successful capture would have resulted in an elongated comet-like orbit — not circular."[8]

3. The moon condenses out from a cloud of dust

This is the view that the Earth–moon system formed from an accretion of dust or gravel-size particles. This scenario hits enormous difficulties, because the initial conditions for gravity to begin to have an effect on such a cloud of gas, dust, or particles require that the density of the cloud must be large. And no cloud will start off dense. This effectively is assuming the solution, rather than solving the problem of gravity bringing everything together in the first place.[9] As with the first scenario, this would not explain the low iron content of the moon.

4. The moon formed by a collision with Earth

This hypothetical idea is believed by many today and is touted strongly by the media,[10] but it has some major problems. It is suggested a larger Mars-sized object hit Earth, or alternatively a smaller object hit a fast spinning proto-Earth. However, both scenarios then lead to too much angular momentum (spinning energy) for both Earth and the moon.[11] The rare titanium isotopes in moon rock should not be the same as those found on Earth and, furthermore, the moon has a weak magnetic field,[12] which is totally inconsistent with a billions of years history for all these four scenarios. It should all have disappeared.

The moon is indeed created

Everything is really showing that the moon is created and that its position, size, almost circular orbit, and the fact that it is approximately 400 times smaller than the sun but is 400 times closer than the sun (and thus total eclipses can take place), are all indicating that the moon is designed perfectly for its function. As Lissauer admitted in an article in *Nature* in 1997: "It's not easy to make the moon!"[13] If man cannot, by natural processes, explain the closest celestial object without creation, how can he speak of the most distant objects with any authority?

Other design features of the moon

The reflectiveness of the moon is exceptionally low (this is called the "albedo"), with a value slightly brighter than that of worn asphalt. Nevertheless, it is the brightest object in the sky after the sun, just as was commanded in Genesis 1:16: "God made two great lights; the greater light to rule the day, and

8 J. Sarfati, *Creation Ex Nihilo* 20(4):36–39, 1998.

9 R.L. Dickman, "Bok Globules," *Scientific American*, 236(6), 66–81,1977.

10 B. Cox, "Forces of Nature with Brian Cox — 2 Somewhere in Spacetime," BBC 1, first shown July 2016.

11 D. Clery, "Impact Theory Gets Whacked," *Science* 342 (6155): 183–185. 2013. See also B. Thomas, "Impact Theory of Moon's Origin Fails," ICR *Creation Science* update, October 28, 2013. The isotopes of titanium found in rocks from the moon are similar to those on Earth. The impact theory predicts these should be different.

12 B. Thomas, "The Moon's Latest Magnetic Mysteries," ICR *Creation Science* Update, June 7, 2013.

13 J.J. Lissauer, "It's Not Easy to Make the Moon," *Nature* 389 (6649), 327–328, 1997.

the lesser light to rule the night." The moon has a cycle of 28 days and is thus used, as Genesis states, for signs and for seasons, in particular the lunar month. Though astrology and worship of the moon is wrong, it is noteworthy that many cultures, such as the Mayans of South America, could predict with great accuracy not only eclipses of the sun, but eclipses of the moon.

6 Earthrise as seen from lunar module Apollo 8 on December 24, 1968. This has become an iconic image and was taken on the same day that crew member Bill Anders (followed by Jim Lovell and Captain Frank Borman) read the opening verse of Genesis 1: "In the beginning God created the heavens and the earth. . . ." Sadly, many seeing this image of Earth from the moon have not worshiped the Creator; they have instead worshiped Earth and the creature and have ignored the Lord who made all.

The Planets

The planets and their moons bring strong evidence of the youthfulness of the solar system.

1 The terrestrial planets to scale of our solar system: Mercury (top), Venus (left), Earth (right), and Mars (bottom)

One of the most interesting facts about the solar system is that in terms of its total mass, the sun has 99.85% and the planets and their moons the remainder. However, the angular momentum (that is, loosely the amount of spin — the force times angular velocity around the center) of the solar system rather than primarily being in the sun, as one would expect, is 97% in the planets and the moons. In fact, 60% of all the angualr momentum of the solar system is in Jupiter. This shows that the idea of the planets being formed from a condensing dust/vapor cloud around the sun is not coherent, since the physics of such an imaginary development would lead to the bulk of the angular momentum being close to the center of rotation where the greater mass concentration is. In other words, the planets and moons are going around faster than one would expect if natural causes were flinging out material from a rotating nebula with the sun's period of rotation now being only 25 days. Or to put it another way, the sun is not spinning fast enough to be consistent with this hypothesis of planets being formed from an initial nebula.

In this section we consider the hot inner planets (figure 1) Mercury and Venus.

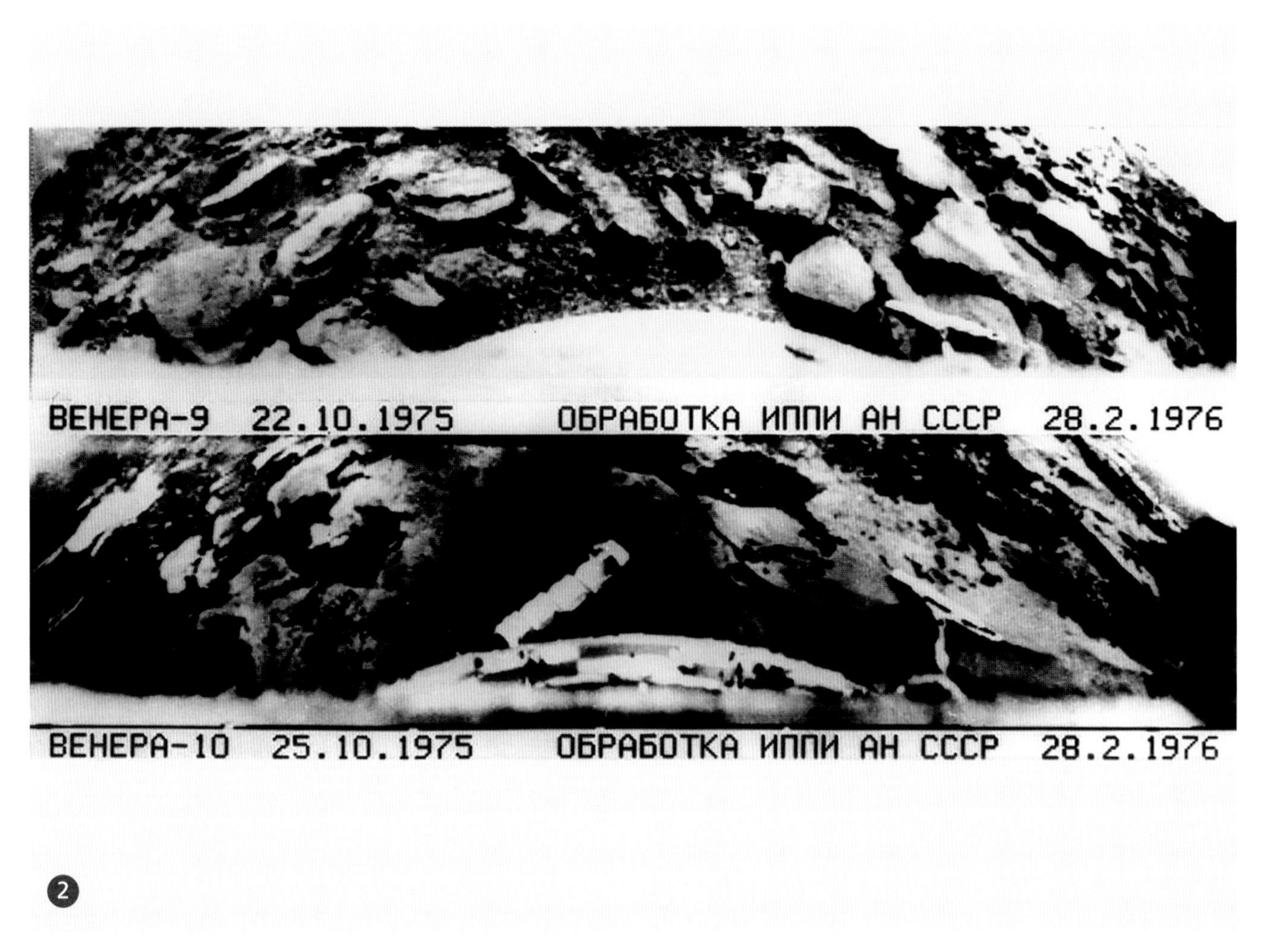

2 Surface of Venus from the Russian probes Venera 9 (October 22, 1975), and Venera 10 (October 25, 1975)

The hot inner planets — Mercury and Venus

Two diverse worlds: one with no atmosphere at all and one with a suffocating atmosphere of sulphur dioxide.

Mercury

Mercury is the closest planet to the sun and small — just a little larger than our moon. It has an orbital period of 88 days, and because of its closeness to the sun, its temperature varies greatly between a day temperature of 427°C (800°F) and a night temperature of -173°C (-280°F). However, at the poles, because the planet barely has any tilt in its axis, the temperature is always cold at -93°C (-136°F). It has an odd movement in its orbit around the sun whereby it rotates on its axis (with respect to the stars) exactly three times for every two orbits it makes around the sun. This means that an observer on Mercury would only work through one solar day every two Mercury years (that is every 176 Earth days = 2 Mercury orbits of 88 days round the sun). The path of Mercury is much more elliptical than the other planets and this leads to some unusual phenomena. At some points in Mercury's orbit its angular orbital speed (that is how fast it goes round the sun) exceeds its angular rotational velocity (that is how fast it spins) which means that an observer on Mercury would see the sun for part of the Mercury

3 The surface of Venus from the Russian probe Venera 13 (March 1, 1982)

"day" actually go backwards — it would perform a very odd path in the sky.[14]

Though Mercury is bright, it is very difficult to see from Earth because it is so close to the sun, and can only really be observed at dawn or dusk near Earth's horizon on a clear morning or evening. Its appearance is very cratered, rather like our own moon, which indicates that in the past there has been a major period or periods of collisions across the solar system. For those who propose an evolutionary explanation for the origin of the planets, with the planets emerging out of a solar nebula, Mercury is one of the most difficult to show its origin this way, since any early nebula with a contracting hot sun would vaporize most of the present Mercury surface rock. The surface rock and inner core of Mercury is dense, suggesting that its content is rich in iron, more so than any other planet in our solar system.[15]

Mercury has a strong magnetic field[16] which points to the youth of the solar system — this is a disturbing fact for those who believe the solar system is billions of years old. A small planet like Mercury should not be able to maintain a magnetic field for very long. Magnetic fields are discussed later in this chapter .

14 J. Lisle, "The Solar System: Mercury," *Acts & Facts*. 42 (8): 10–12, 2013.

15 R.G. Strom and A.L. Sprague, *Exploring Mercury: the Iron Planet* (Springer, 2003).

16 https://en.wikipedia.org/wiki/Mercury_(planet). Accessed July 2016.

Venus

Venus is almost a twin of Earth in size and is the second planet out from the sun and 26 million miles closer to the sun. Other than the moon, it is the brightest object in the night sky and can easily be observed on a clear evening or morning as it follows the sun from our perspective. Venus orbits the sun every 7.4 months, but its "day" (rotation period) is *longer* than its orbit time and is 8 months! The most remarkable fact is that Venus rotates backward — the sun rises in the west on Venus and sets in the east. That is, viewed from above (with Earth's North Pole in view) the planets go round the sun in an anticlockwise direction and all the planets rotate on their axes also in an anticlockwise sense — except Venus, which rotates in a clockwise direction with an axis tilted only 3 degrees to the ecliptic plane (the plane of rotation of the planets). The evolutionary nebular hypothesis, with planets condensing out of a rotating dust cloud, cannot account for such an odd finding since it would predict all the planets rotating in the same direction at about the same rate. The backward rotation also has the strange effect of reducing the perceived "day" by an observer on Venus to 3.8 months, which is less than the Mercury "day."

Though similar in size to Earth, it is very different in terms of its temperature and atmosphere.[17] The atmosphere is largely made up of carbon dioxide,

17 J. Lisle, "The Solar System: Venus," *Acts & Facts*. 42 (9): 10–12, 2013.

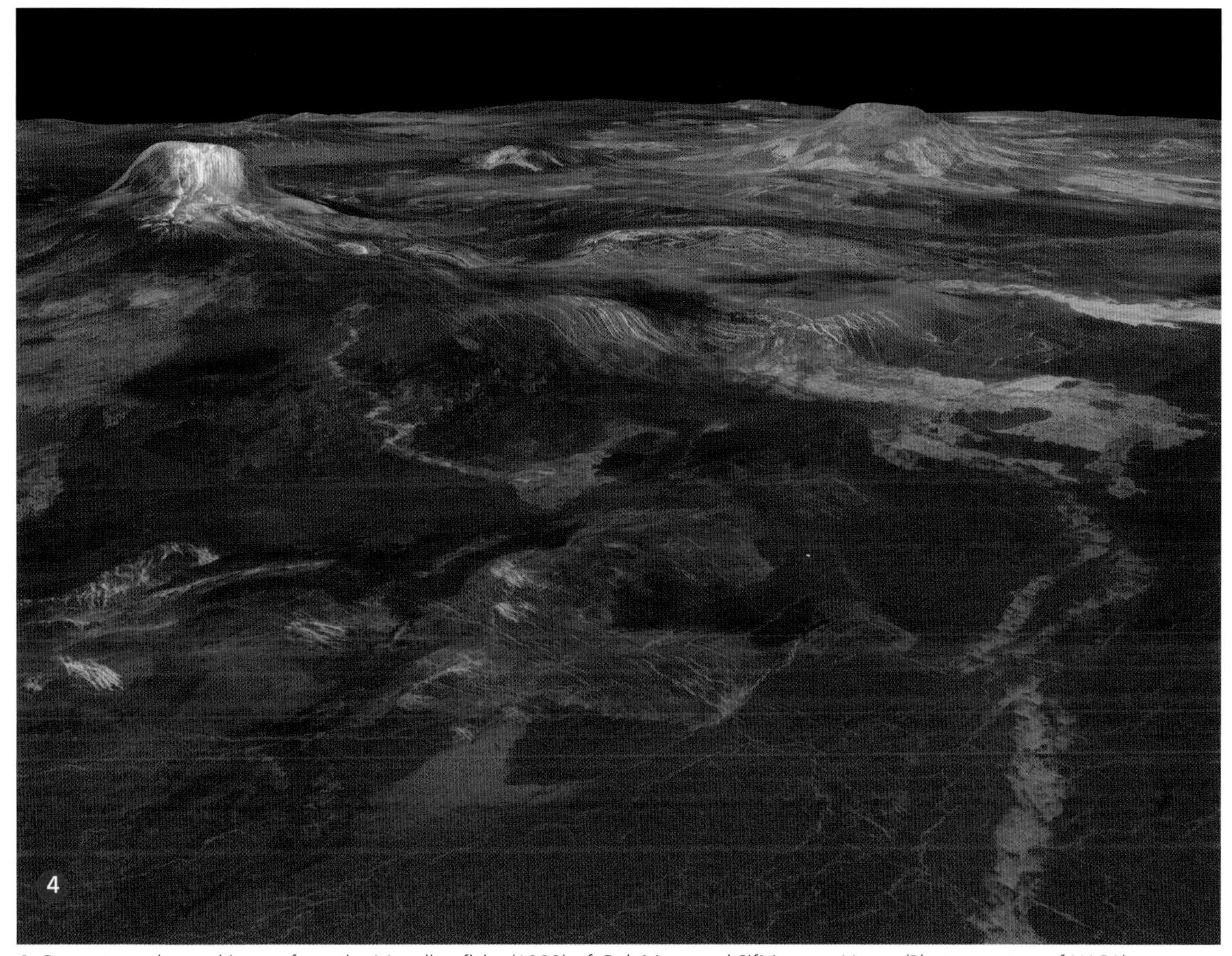

4 Computer enhanced image from the Magellan flyby (1992) of GulaMons and SifMons on Venus *(Photo courtesy of NASA)*

with thick opaque clouds of sulphur dioxide and sulphuric acid containing traces of hydrochloric acid, bromine, and iodine.[18] The thick clouds mean that the surface cannot be observed from Earth. The atmosphere acts like a blanket and produces a strong greenhouse effect, raising the temperature to a scorching 485°C (905°F) at the surface and producing a pressure 90 times that of Earth's atmosphere. Probes from Russia successfully landed on Venus in 1975 and 1982 (see figures 2 and 3) and sent the first successful pictures back to Earth. These probes were not designed to last longer than about 30 minutes in such hostile conditions. In actuality, Venera 9 lasted 53 minutes, Venera 10 lasted 65 mins and Venera 13 lasted 127 minutes, so considerable knowledge was gained by the Venera missions and by the NASA Magellan flyby (figure 4).

Unlike Mercury, Venus has a weak magnetic field, much weaker than Earth; and Venus, like our moon, goes through phases which will alter the apparent brightness that we see in the morning or evening sky. We only see the full sphere of Venus when it is at its farthest distance from Earth. As it comes closer to us and begins to overtake us (every 19 months) it is larger, but with only part illumination. Its striking appearance in the evening or morning sky gives great glory to the Lord who has formed it.

18 https://en.wikipedia.org/wiki/Venera_9; https://en.wikipedia.org/wiki/Venera_10; https://en.wikipedia.org/wiki/Venera_13. All accessed July 2016.

Mars

A world where water once flowed . . .

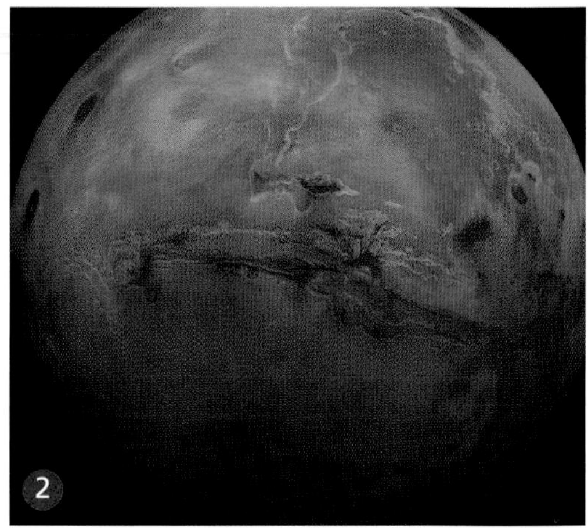

1 Curiosity on Mars taking a selfie (October 6, 2015) near Mount Sharp in the Gale Crater. The evidence of mudstone layering is striking. *(Courtesy NASA/JPL-Caltech/MSSS)*

2 Mars showing the Valles Marineris — a large scar which is caused by a deep crack in the Martian crust. *Valles Marineris* is 2,500 miles long and reaches depths of up to 4 miles. September 7, 2013. *(Courtesy NASA/JPL-Caltech)*

With a thin atmosphere of carbon dioxide, Mars has a surface that can readily be observed through telescopes on Earth. With the highly successful landing on Mars by the NASA Curiosity Rover (August 2012),[19] detailed knowledge of this red planet has now been obtained — many clear color photographs are being sent at regular intervals as Curiosity explores the region (see figures 1 and 2). Eventually a manned mission is planned. Mars is just over half the size of Earth and has about 40% of the gravitational pull of Earth. The Martian year is 1.9 Earth years and the rotation of the planet gives it a day very similar to that on Earth, of 24 hours 37 minutes. Mercury and Venus have no moons, Mars has two — Phobos and Deimos.

Mars has an axial tilt of 25.2 degrees, which is similar to Earth's tilt of 23.5 degrees and the Martian temperatures can vary from -85°C to 7°C. Its orbit is quite elliptical, and because its orbit is farther out than Earth, there comes a point when it is said to be in opposition — that is when the sun is directly behind illuminating the planet (in this case Mars) at its closest position to Earth and opposite the sun. The opposition of Mars only happens every 2.1 years. At such times, Mars can appear 7 times larger and 50 times brighter than it does when on the far side of the sun.[20]

19 This brilliant engineering feat enabled Curiosity to land without mishap at Aeolis Mons or Mount Sharp in the Gale crater just south of the equator of Mars.

20 J. Lisle, "The Solar System: Mars," *Acts & Facts*. 42 (11), Nov. 2013; http://www.icr.org/article/7721.

3 Curiosity shows evidence of sedimentary rocks at Whale Rock at base of Mount Sharp in Gale crater. The scale of the photo is approximately 1.5 meters. November 2, 2014. *(Courtesy NASA/JPL-Caltech/MSSS)*

The terrain of Mars is desert-like with a rocky surface that includes mountains, volcanoes, canyons and many dry riverbeds and deltas. Mars has the largest volcano and the second-highest known mountain in the solar system, Olympus Mons. Many think that there was liquid water at one time on Mars[21] which caused the riverbeds, although flowing liquid water has not been found today. In fact, with such a thin atmosphere, ice when it melts would simply sublime to a vapor (sublimation means that it goes straight to being a vapor and does not go first to a liquid because of the very low pressure). Ice has been found in the soil,[22] and the detailed photographs returned from Mars by the Curiosity Rover since August 2012 show clearly that Mars originally had flowing water but then, due to the low gravity of Mars, this was gradually lost. The fact that there is water ice near the poles, a deep crack in the Martian crust (*Valles Marineris* — see figure 2), and the more recent findings of mudstones and rocky outcrops such as those illustrated in figures 1 and 3 strongly suggest this. Some even believe that Mars had catastrophic, planet-scale flooding — on a planet with no liquid water today — though the same scientists deny this happened on Earth which has 70% of its surface covered with rocks laid down with water from a worldwide Flood, and for which there is abundant evidence!

There is no evidence that Mars has, or once had, a significant magnetic field, though some observations show that parts of the planet's crust have been magnetized in the past, with some evidence of alternating polarity reversals. With such a cold temperature on Mars one would expect that the magnetic field would be strong, so the lack of a structured magnetic field was a surprise to many scientists.

The two moons of Mars

These moons are very small. Phobos, the inner moon, is approximately just 10 miles wide and has a circular orbit only 3,700 miles above the surface of Mars, orbiting Mars in just 7 hours and 39 minutes. Deimos is 8 miles wide and takes 30 hours to complete its orbit. Neither moon is spherical (round), since their gravity is not strong enough to pull them into a spherical shape. These moons represent an enigma to evolutionary scenarios since they have near perfect circular orbits. The prevalent idea is that Mars captured these two moons as asteroids, but if that were the case, captured asteroids usually have highly elliptical orbits.

21 J.P. Grotzinger et al., "Stratigraphy and Sedimentology of a Dry to Wet Colian Depositional System, Burns Formation, Meridiani Planum," *Earth and Planetary Science Letters* 240 (1): 11–72, November 25, 2005.

22 "NASA Spacecraft Confirms Martian Water, Mission Extended," *Science @ NASA*. July 31, 2008. See http://www.nasa.gov/mission_pages/phoenix/news/phoenix-20080731.html, accessed July 2016; also see Gina Anderson, ed. (September 28, 2015), "NASA Confirms Evidence That Liquid Water Flows on Today's Mars" (press release), NASA, https://www.nasa.gov/press-release/nasa-confirms-evidence-that-liquid-water-flows-on-today-s-mars. Accessed July 2016.

Jupiter

The largest planet of all — the vacuum cleaner of the solar system!

1 Jupiter photo taken from the Hubble Telescope. (*Courtesy NASA/ESA)*

Jupiter (figure 1) is one of the most magnificent sights in the universe. It is the largest planet in the solar system and one of the brightest objects visible in the heavens after Venus. Jupiter is 11 times wider than Earth and 318 times more massive than Earth, although on average 500 million miles farther away. It takes 12 Earth years to orbit the sun, rotates on its axis every 10 hours, and the axis is only tilted 3 degrees to the ecliptic plane.

In composition, Jupiter resembles a star in that it is mostly hydrogen and helium. Under pressure, the hydrogen becomes an electrically conducting fluid, and it is probable that this "metallic hydrogen" is the source of its very strong magnetic field, which is 14 times as strong as that of Earth.[23] Other than the sun, it has the strongest magnetic field in the solar system.

Jupiter has thick gases swirling round its huge globe. Most of the visible cloud-tops contain ammonia and hydrogen sulphide, and Jupiter's "stripes" are created by strong east-west winds. One noticeable part of Jupiter is what is called the Great Red Spot, which is a giant storm vortex twice as wide as Earth!

23 Jupiter's magnetic field ranges from 4.2 gauss (0.42 mT) at the equator to 10–14 gauss (1.0–1.4 mT — that is a thousandth of a Tesla, which is the unit used in magnetic fields) at the poles.

2 Jupiter's Galilean moons — Io, Europa, Ganymede, and Callisto (in order of increasing distance from Jupiter)

Jupiter's mass is 2.5 times that of all the other planets in the solar system combined and radiates more heat than it receives from the sun; the amount of heat produced inside it is similar to the total solar radiation it receives.[24] This additional heat is generated by the contraction of Jupiter by about 0.8 inch each year.[25] It is probable that below the inner metallic hydrogen there is a cold solid core, but the size of this is not yet known. The large magnetic field and the inner heating strongly suggest that Jupiter is not millions of years old because both the magnetic field strength and the heat dissipate over time. This is consistent with a few thousand years of existence.

The moons of Jupiter

Amazingly, Jupiter has at least 67 moons! That went up to 79 in July 2018, with one of the extra 12 going against the direction of its neighbors.[26] The four largest moons (figure 2) are called the Galilean moons, named after their discoverer Galileo Galilei in 1610. Ganymede is the largest and has a diameter greater than that of the planet Mercury. Jupiter and its moons are like a miniature solar system with the inner moons orbiting faster than the others. Io orbits Jupiter every 1.77 days, and the orbital periods of Europa, Ganymede, and Callisto are 3.55 days, 7.15 days, and 16.7 days respectively. So that the three inner moons are in what is called "orbital resonance" — Europa's period is twice that of Io, and Ganymede's period is twice that of Europa.[27] Io is the most active moon, receiving gravitational stretching every time it passes between Jupiter and Europa which causes volcanic upheaval on Io, with volcanoes (some taller than Everest) spewing out sulphur and sulphur dioxide up to 300 miles above the surface. Of the other 63 moons, 52 of these are in retrograde orbits (i.e., orbiting in a clockwise orbit, opposite to the counterclockwise spin of Jupiter), but in such a way that they are all in stable orbits and they do not collide.

Because of its massive gravitational pull, all probes sent up in our solar system have to take account of Jupiter's pull. It also pulls away any stray objects which could endanger Earth, and in that sense acts like a vacuum cleaner in the solar system. An example is comet Shoemaker–Levy 9 that collided with Jupiter in July 1994.

24 L.T. Elkins-Tanton, *Jupiter and Saturn* (New York: Chelsea House, 2006).

25 T. Guillot, D.J. Stevenson, W.B. Hubbard, D. Saumon, chapter 3: "The Interior of Jupiter," in F. Bagenal, T.E. Dowling, and W.B. McKinnon, *Jupiter: The Planet, Satellites and Magnetosphere* (Cambridge University Press, 2004).

26 L. Grossman, "Jupiter Has 12 More Moons than We Knew about — and One Is Bizarre," *ScienceNews*, July 17, 2018; see also A. Witze, Nature 559, News and Comment, 312-313, July 17, 2018.

27 J. Lisle, "The Solar System: Jupiter," *Acts & Facts*. 42 (12), December 2013; http://www.icr.org/article/7834.

Saturn

Saturn's rings make this the most beautiful object in the night sky.

1 Saturn from Cassini 2008/2009

Saturn, with its rings (figure 1), is arguably the most awesome object to observe in the night sky. When first viewed through a telescope, Saturn is breathtaking in majesty and has been compared to a piece of celestial jewelry.[28] Most of us will only see it climb high in the night sky over us once or twice in our lifetime, because it takes nearly 30 years to go round the sun. It is slightly smaller than Jupiter and nine times the size of Earth. Saturn is 890 million miles away from the sun and, like Jupiter, it is a gas giant made up mainly of hydrogen and helium. It is also believed to have an icy core like Jupiter and has a region of metallic hydrogen and helium linked to a magnetic field which is slightly weaker than Earth's and 20 times weaker than Jupiter's.[29] Saturn's rings span 170,000 miles across and are less than 1 mile thick. Because the axis of rotation of Saturn is tilted at 26.7 degrees, this means that we see the rings at different angles as Saturn gradually makes its long orbit round the sun.

Saturn has 62 moons that are known (see figure 2). Its biggest moon is Titan, which is second only to Ganymede, the largest moon of Jupiter, and 48% bigger than Earth's moon. Titan has an atmosphere mainly made of nitrogen and is colored orange due to the presence of hydrocarbons. This is unique for any moon in the solar system.

There are other moons, some of which are intriguing, such as the small moon Enceladus (diameter 313 miles) which was observed by the probe Cassini to have jets of water vapor and ice coming out of its south pole (figure 3). As stated by Lisle,[28] this indicates "that Enceladus has significant internal heat. This is problematic for those who believe that the solar system is billions of years old because if the moon were really that old, that heat should have escaped long ago. Unlike Jupiter's moon Io,

28 J. Lisle, "The Solar System: Saturn," *Acts & Facts*. 43 (1), 2014, http://www.icr.org/article/7860.

29 Saturn's magnetic field is 0.2 gauss (20 μT that is 20 millionths of a Tesla).

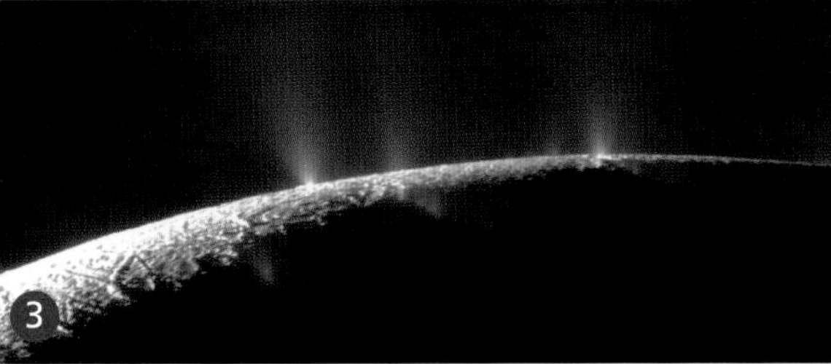

2 Major moons of Saturn *(Courtesy of NASA/JPL-Caltech/Space Science/Kevin M. Gill)*

3 Cassini probe sees jets of water vapor and ice from Enceladus *(Courtesy of NASA/JPL-Caltech/Space Science Institute, November 21, 2009)*

4, 5 Iapetus with its strange ridge along the equator making it look like a walnut

Enceladus does not experience enough gravitational tugging to regenerate its internal heat." This evidence fits well with the view that the age of the solar system and indeed the universe, is measured in thousands of years, not billions.

Iapetus (mean diameter 913 miles) is the third largest moon of Saturn after Titan and Rhea. Iapetus has a ridge running all along its equator (see figures 4 and 5). The ridge is 12.5 miles wide and 8 miles high, with peaks as much as 12.5 miles high, so that this moon looks like a walnut! The ridge and its formation is a mystery not yet resolved.

Some moons have prograde orbits (anticlockwise) and some retrograde (clockwise), and there are some very small moons within the rings which shepherd other moonlets so that they gravitationally deflect any wayward ones back into the rings! There are even two moons which swap their orbits and so avoid a collision! The more that is discovered about what is effectively a mini solar system round Saturn, the more one stands in awe of the splendor and complexity of what God has made.

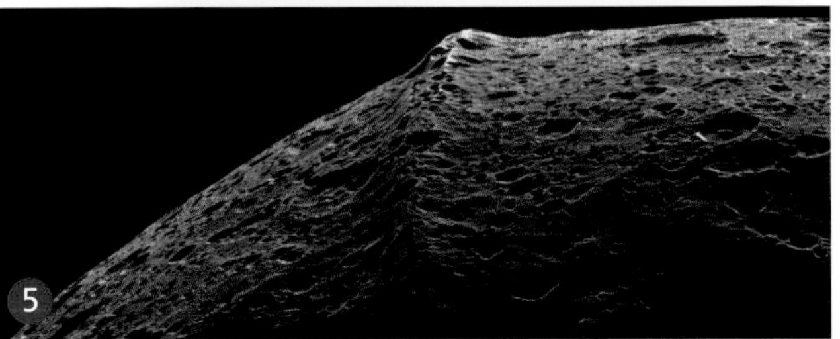

Uranus and Neptune

Secret worlds at vast distances away from the sun.

1 Montage of Uranus and its five main moons. The moons, from largest to smallest, are Ariel, Miranda, Titania, Oberon, and Umbriel. *(Courtesy of NASA/JPL)*

Uranus

Uranus (figure 1) is another gas world 4 times the size of Earth, and 1,790 million miles from the sun (19 times the orbit of Earth), taking 84 years to go around the sun. Mainly hydrogen, helium, and a little methane, it is thought to have a small iron-nickel core, and it has 27 moons with 5 main ones (see figure 1). Uranus was discovered by Sir William Herschel in 1781, and only in 1977 was it also discovered to have rings like Saturn, although fewer and thinner.

Uranus has two most remarkable features. One is that it rotates on its side so that its axial tilt (the tilt of its axis of rotation to the ecliptic plane) is 97.8 degrees (see figure 2). This means that its north pole of rotation is actually slightly pointing downward to the orbital plane of the planets! Therefore, at certain times in the Uranus "year" an observer would see the sun doing a circle in the sky and at other times no sun would be visible at all. The other strange feature is that the magnetic field of Uranus is not aligned to its rotational axis and is displaced to one side. Magnetic field strengths decay, so all magnetic fields in the solar system should be absent if the solar system were billions of years old. Secular evolutionary ideas propose dynamo heat theories to keep the mantle moving and create the magnetic fields. However, Uranus lacks any measurable internal heat, which would be essential to keep its magnetic field going over billions of years.[30] Such proposals are really an invention to avoid the real implication that Uranus is not ancient.

In 1984, Russ Humphreys, using the free decay model for planetary magnetic fields, predicted the field strength of Uranus by assuming it had been

30 J. Lisle, "The Solar System: Uranus," *Acts & Facts*. 43 (2), February 2014, http://www.icr.org/article/7885.

slowly decaying for 6,000 years.[31] As Voyager 2 flew by Uranus in January 1986, the field strength was found to be close to this value. This is an excellent confirmation of the creation-based understanding of the relatively young solar system.

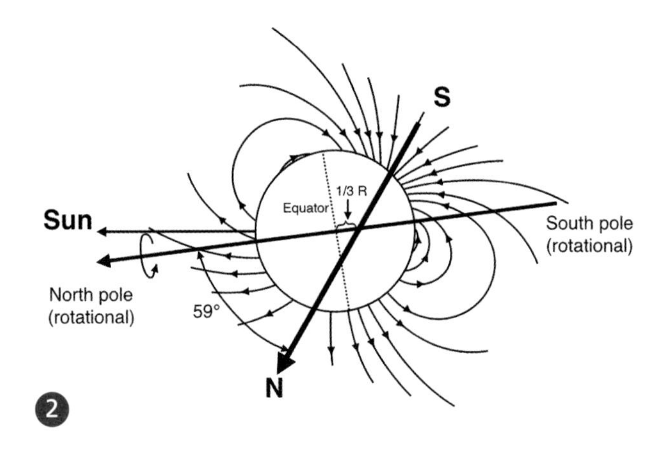

2 Uranus rolls on its side and its magnetic field is not aligned to its axis of rotation, plus the magnetic field axis *is shifted to one side*. Neptune also has a magnetic field tilted and shifted in a similar way.

Neptune

With an orbital distance of approximately 2.8 billion miles — which is more than 30 times that of Earth — Neptune (figure 3) is very difficult to detect. However, because of the discrepancies in the orbit of Uranus, this large planet, similar in size to Uranus, was known to exist before its actual observation in 1846. The tilt of its axis to the ecliptic is 28 degrees, similar to most of the planets, and it takes 165 years to orbit the sun, so it has actually only completed 36 orbits since creation. Like Uranus, its composition is mainly hydrogen and helium with some methane. It also has some small rings but much fainter than those of Uranus. They were observed by Voyager 2 as it passed Neptune in 1989.

Neptune has a number of moons, with one major one called Triton. Lisle notes that Triton's orbit is retrograde, that is, opposite to the direction of spin of Neptune.[32] Furthermore, Lisle comments on Triton's orbit: "Large moons generally orbit in the plane of their planet's equator, but Triton breaks this rule as well and orbits at an angle of 23 degrees." Neptune, like Uranus, has a magnetic field totally misaligned with its rotational axis — it is tilted at 47 degrees and displaced by as much as half the planet's radius. However, unlike Uranus, although this remote planet does have significant internal heat, it still has a similar strength magnetic field to Uranus. This again is consistent with the free decay model of planetary magnetic fields and the planets not being billions of years old.

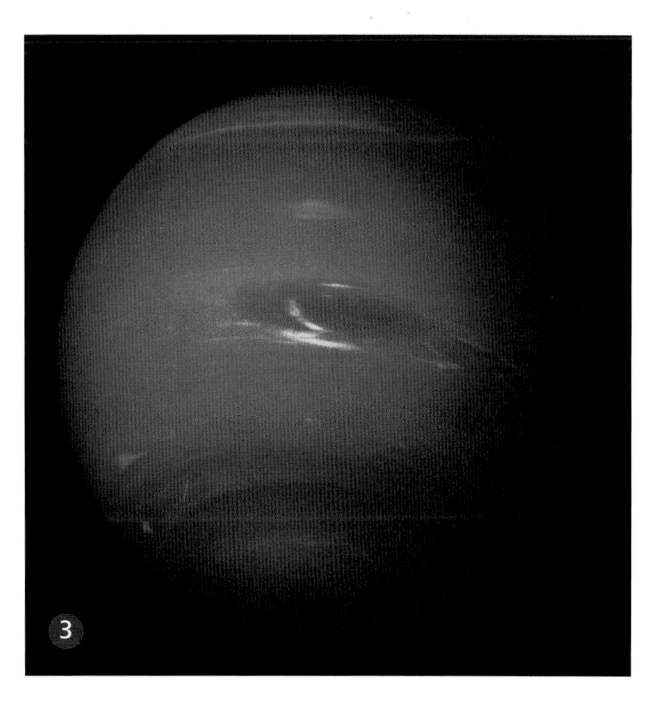

3 Neptune as viewed by Voyager 2 in August 1989. *(Courtesy of NASA/JPL)*

31 D.R. Humphreys, "The Creation of Planetary Magnetic Fields," *Creation Research Society Quarterly Journal*. 21 (3), 1984; http://www.creationresearch.org/crsq/articles/21/21_3/21_3.html.

32 J. Lisle, "The Solar System : Neptune," *Acts & Facts*. 43 (3). March 2014; http://www.icr.org/article/7906.

Pluto, Planet Nine, and More

Pluto is relegated to a dwarf planet, but offers intriguing evidence of more we haven't yet seen.

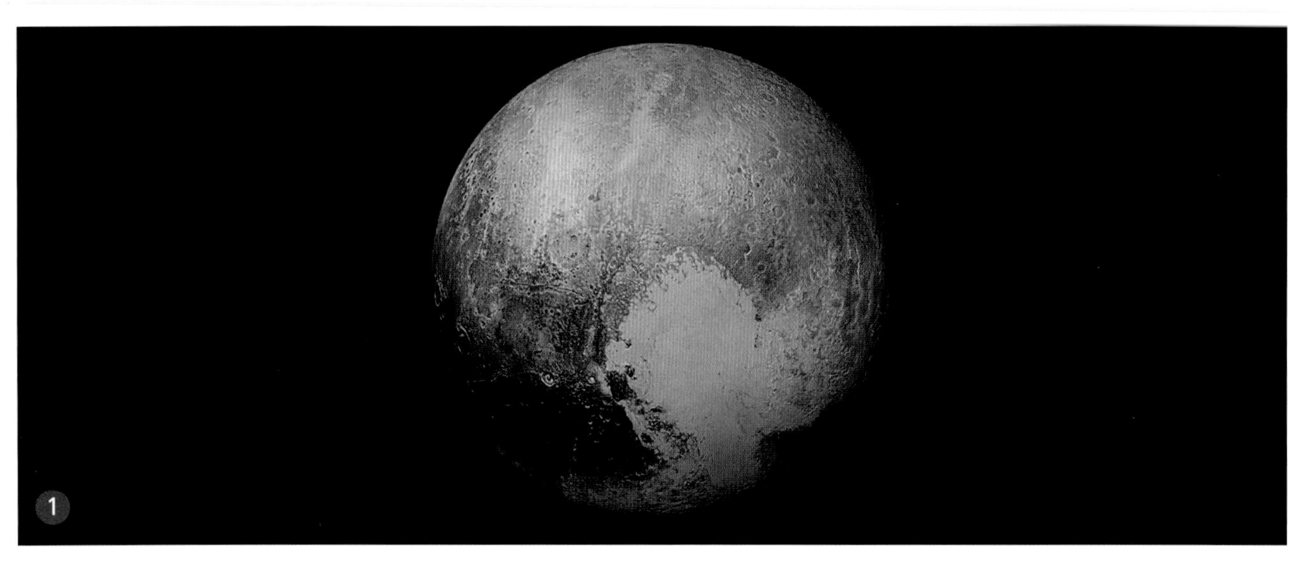

1 Pluto from New Horizons probe July 2015. *(Courtesy of NASA/JPL)*

Pluto, discovered in 1930, is no longer strictly classified as a planet because it is small with a diameter of 1,430 miles or one-fifth the size of Earth, and its elliptical orbit comes within that of Neptune's for part of its very slow 248 year travel around the sun. Its orbit is highly eccentric and the plane of its orbit is at an angle of 17 degrees to all the other orbiting planets, with its axial tilt of 120 degrees. Therefore, like Uranus, the dwarf planet Pluto rolls on its side. In July 2015 a probe called New Horizons (originally launched back in January 2006) finally reached its destination and began to send back spectacular images (figure 1) of this very distant world which is a chilly 40 K (that is -233° C) and is made up of frozen nitrogen, methane, and carbon monoxide. On average, Pluto is 3,670 million miles from the sun. At its closest, it is 30 times the distance from the sun as Earth, and at its farthest it is 40 times our distance.

Pluto has 5 known moons and its largest is called Charon, with a diameter just over half of Pluto's. Both Pluto and Charon are locked into each other's orbits, so Charon keeps the same side pointed toward Pluto as it revolves and Pluto also does the same with Charon. Their joint orbital period is 6.4 days, so really they are acting as a dwarf planet pair.[33]

Pluto is one of many Trans Neptune Objects (TNOs) which have now been found. Sometimes these are referred to as Kuiper Belt Objects (KBOs). These are smaller than planets, and cross Neptune's orbit with very eccentric orbits, but none have been found as small as comets. One similar in size to Pluto, called Eris, was found in 2005 and is at an orbital average distance of 53 times Earth's distance from the sun, taking 558 years to complete one orbit. Another is Sedna, which has a highly eccentric orbit, such that its distance closest to the sun is 76 times the distance of Earth to the sun, and at its farthest it

33 J. Lisle, "The Solar System: Pluto," *Acts & Facts*. 43 (4), April 2014; http://www.icr.org/article/8019.

reaches 937 times this distance. It is approximately 618 miles in diameter and would take 11,400 years to go around the sun — therefore it has only completed half an orbit since creation!

Some scientists are suggesting that the behavior of some of the TNOs is indicative of a large object called Planet Nine yet to be discovered. Though not yet confirmed, the evidence is mounting that there could be a planet ten times the size of Earth at a distance of 450 times the distance of Earth to the sun.[34]

Asteroids

Asteroids are in a belt of objects which lie in a region between Mars and Jupiter. The largest of these is Ceres (figure 2), which is the only asteroid that is massive enough to become rounded under its own gravity. It is approximately 587 miles in diameter, and is really a dwarf planet and the largest object out of hundreds within the asteroid belt. It orbits the sun every 4.6 years. The NASA spacecraft Dawn delivered close-up pictures of this asteroid / dwarf planet in May 2015.

Comets

Comets are distinguished from asteroids since they are much smaller. Asteroids are composed of rock, whereas comets are essentially made of ice and dirt.[35] Since comets mainly consist of ice, they actually lose a bit of their small mass with evaporation every time they come near the sun in their very eccentric (highly elliptical) orbits. The last major one was Hale Bopp in 1997, which was a brilliant sight (figure 3) and its path was almost at 90 degrees to the ecliptic plane. The evaporation is what gives the famous tail of comets, which are very impressive in the night sky. In August 2014, the Rosetta spacecraft actually landed on Comet 67P (figure 4), which is approximately a tiny 2.5 miles long! The fact that we still have short period comets which take less than 100 years to go round the sun is, again, consistent with a young solar system.

34 K. Batygin and M. Brown, "Pathway to Planet Nine," *Physics World*, 29(7), 28–31 July 2016.

35 J. Lisle, "The Solar System: Asteroids and Comets," *Acts & Facts*. 43 (5), May 2016; http://www.icr.org/article/8045.

2 The dwarf planet / asteroid Ceres. Taken from NASA probe "Dawn." *(Courtesy of NASA / JPL-Caltech /UCLA /MPS /DLR /IDA / Justin Cowart)*

3 Hale Bopp comet viewed from Croatia. *(Photo by Philipp Salzgeber, March 1997)*

4 Comet 67P photographed by the Rosetta mission where the Rosetta lander actually landed on the comet in August 2014. *(Photo by ESA/Rosetta/MPS)*

The Stars

"The heavens declare the glory of God" Psalm 19:1.

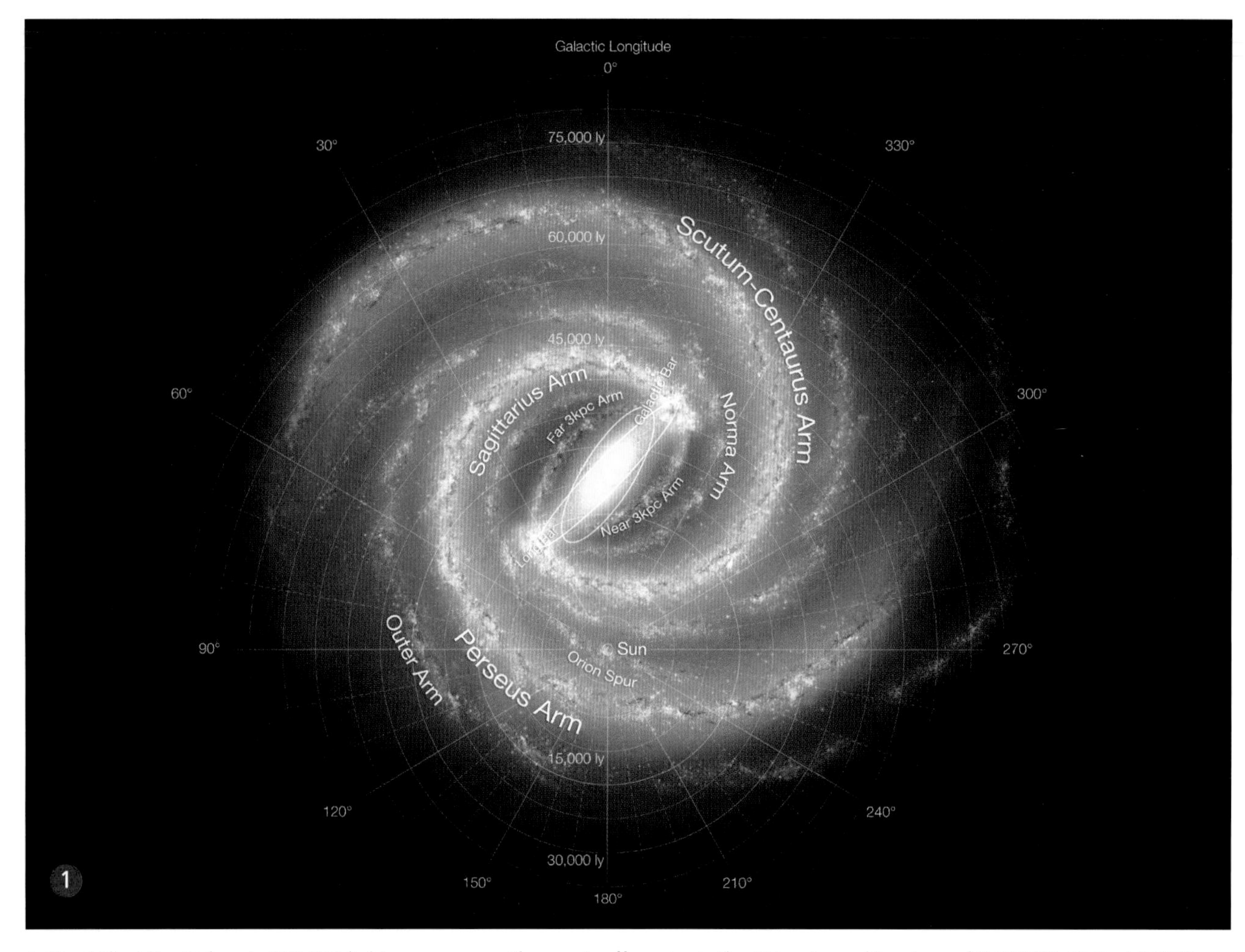

1 The Milky Way Galaxy is 100,000 light years across. The sun is off-center in the Orion Spur. *(Courtesy of NASA/JPL-Caltech/ESO/R. Hurt)*

The nearest star beyond our solar system is Alpha Centauri, which is 4.37 light years away from the sun. This means that distances are now so large that it is better to use a different "ruler." A light year is the distance covered by light in one year, which is 5,879 billion miles — written as 5.879 x 10^{12} miles. The numbers become difficult to appreciate, but this means a light year is nearly 6 million million miles, which is 63,215 times the distance of Earth to the sun. We call the distance of Earth to the sun an AU (Astronomical Unit — 93 million miles). To set this in context, the solar system is approximately 10 light hours across which is 6,706,000,000 miles or 72 AU. Alpha Centauri is 63,215 x 4.37 AU, which is 276,250 AU. These distances are so big as to be unimaginable, and yet this simply takes us to the nearest star in our own galaxy (the Milky Way)! We have only just begun to paddle in the vastness of the ocean of space!

The Milky Way

The Milky Way galaxy itself measures a distance across of 100,000 light years — that is, light takes 100,000 years to cross the galaxy at a speed of 186,282 mi/sec. This distance is equivalent to 6,321,500,000 AU! The sun is on the Orion Spur (figure 1), not too close to the center such that our own galaxy would dominate, and neither too far away such that very few other stars would be visible. The arm of the Milky Way is indeed a dramatic sight as illustrated in figure 2.

There is a design in the location of our sun and its solar system in the Galaxy. This is discussed by Professor Stuart Burgess in his book *He Made the Stars Also* where he shows the design and beauty evident in the colors and formation of nebulae and vast dust clouds visible in a number of star formations.[36]

Other galaxies

There are other galaxies /clusters, such as the Large and Small Magellanic Cloud at 179,000 and 210,000 light years respectively, which are effectively satellite galaxies revolving around the Milky Way. These are all part of what is called the Local

36 S. Burgess, *He Made the Stars Also*, (Day One Publications 2001); see in particular chapter 6, "The Beauty of the Universe."

2 The Milky Way, photographed in October 2010 *(Courtesy of ESO / S. Bruni)*

3 Andromeda Galaxy M31, which is 2.5 million light years distant. *(Photo courtesy of Adam Evans, September 2010)*

Galactic Group. The nearest full Spiral Galaxy to the Milky Way is Andromeda (figure 3) which is 2.5 million light years distant and part of this group. It is estimated that the number of stars in our own galaxy is approximately 200,000 million[37] and that Andromeda may well have more than this. The sheer scale of such numbers and distances is awe inspiring, and the beauty of the spiral and barred type of galaxies is demonstrated in figures 3, 4, and 5.

When you look at the stars on a clear night without light pollution, you begin to appreciate what Abraham in the Old Testament must have thought when the Lord God asked him: "Look now toward heaven, and count the stars if you are able to number them . . ." (Genesis 15:5). The best place to observe the grandeur of the stars is away from light pollution. This author (Andy McIntosh) has had the privilege to see them in Arizona when sleeping out in the open in the Grand Canyon! Their magnificence shouted the truth of Psalm 19:1 "The heavens declare the glory of God."

37 N. Henbest and H. Cooper, *The Guide to the Galaxy* (Cambridge University Press 1994). Also see H. Frommert and C. Kronberg, "The Milky Way Galaxy," SEDS, August 2005; http://messier.seds.org/more/mw.html. Accessed July 2016.

How many galaxies and stars?

How many galaxies are there in the visible universe? Over 2 million have been counted, but there could be as many as 100,000 million, that is 100 billion. This implies that a conservative estimate for the number of stars could be as many as 200,000 million x 100,000 million. Such a number is beyond comprehension — it is 20,000,000,000,000,000,000,000, written in shorthand as 2 x 10^{22} that is 2 with 22 zeroes after it.

If a computer were observing 3 million per second, it would take 211 million years to count all the stars. At 10 million per second, it would still take 63 million years! Yet Psalm 147:4 states that God knows all the stars by name!

Measuring the distance of the stars and galaxies

To measure the distance to a star up to about 500 light years, astronomers use a system called "parallax." This means that when the star is observed 6 months apart, because of the 186 million miles separating the two locations in Earth's orbit around the sun, the star will appear to be in a slightly different angular location. Then geometrical calculations based on the two observation angles can deduce the distance that the star is from Earth.

Up to 30 million light years distant from Earth, astronomers use Cepheid variable stars which are observed in stars close to us, to vary their brightness in a predictable manner. Consequently, when these are observed with variable brightness, but at a much lower level, their distance is deduced from that level. This is rather like car headlights in the distance at night giving an idea as to how far away the car is. The assumption of course is that all Cepheid variable stars have the same basic level of brightness for the same distance.

Beyond 30 million light years, astronomers use what is called the Hubble Law concerning the expansion of the universe. It is observed that other galaxies are generally moving away from us. As they move away, the light from them is redshifted[38] according to their speed. Most galaxies are redshifted though there are a few exceptions.[39] The Hubble Law proposes that the distance is proportional to the speed which is correlated with the redshift of the light received. There are assumptions in this scenario of what is called the Big Bang expansion, the main one being that there is no preferred location that the expansion is taking place from. However, there are tantalizing pieces of evidence which indicate that this assumption is at the very least premature, and that alternative cosmologies (discussed in chapter 7 on pages 126–129) are worthy of consideration. One important evidence is that the redshifts are grouped in clusters close to particular values. This is called "quantization of redshift" and is discussed in Stuart Clark's book *Redshift*[40] and Danny Faulkner's book *Universe by Design*.[41]

4 Pinwheel Galaxy Messier M101 (NGC 5457) *(Courtesy European Space Agency & NASA, February 2006)*

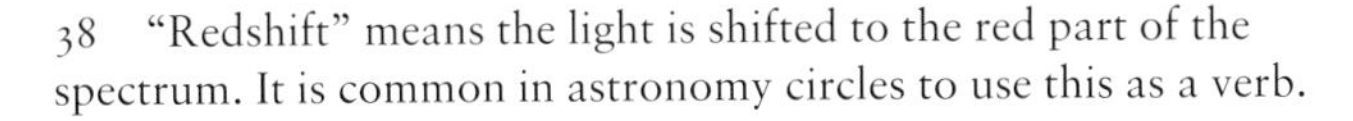

38 "Redshift" means the light is shifted to the red part of the spectrum. It is common in astronomy circles to use this as a verb.

39 J. O'Callaghan, "Apart from Andromeda, Are Any Other Galaxies Moving Towards Us?" February 22, 2003; http://www.spaceanswers.com/deep-space/apart-from-andromeda-are-any-other-galaxies-moving-towards-us/. Accessed July 2016.

40 S. Clark, *Redshift* (University of Hertfordshire Press, 1997), see p 160–165 for discussion of redshift quantization.

41 D. Faulkner, *Universe by Design* (Green Forest, AR: Master Books, 2006), see p 82–83 for discussion of redshift quantization.

5 NGC 1300 — An example of a barred galaxy *(Courtesy NASA, ESA, with Hubble Telescope, Sept 2004)*

If the universe is millions or billions of light years in size, some argue that therefore the light traveling to us is surely showing a very old universe.

Because we can see stars which are millions of light years away, it is thought one is peering into what the universe was like closer to its beginnings, the further out one looks. Since there is so much time involved for the light to get to us, it is argued that this must mean that the universe is much older than 6,000 years. Although this may appear to be an insurmountable argument, it is based on the assumption that the universe evolved from a so-called "Big Bang." However, there are a number of serious problems with the Big Bang theory.

1 Humphreys[42] and Hartnett[43] have proposed different cosmologies which allow for a natural center of mass of the universe near the Milky Way, based on the notion that there is a preferred location, which is our galaxy. Einstein proposed his theory of general relativity, which explained the precession of the long axis of Mercury's orbit,[44] and also predicted the bending of light shown to be true by observations of starlight at an eclipse.[45] He showed that gravity affects light travel significantly, and that with regions of high mass, clocks go slower. With the assumption of much greater mass as the initial universe is expanded, this then leads to clocks going at different rates. These theories are beginning to show possible routes whereby the light travel problem could be resolved in the context of a universe thousands of years old rather than billions.

42 D.R. Humphreys, *Starlight and Time* (Green Forest, AR: New Leaf Publishing. 2010).

43 J. Hartnett, *Starlight, Time and the New Physics* (Creation Book Publishers, LLC, 2007).

44 A. Einstein, "The Foundation of the General Theory of Relativity," Annalen der Physik 49 (7): 769–822, 1916. See https://en.wikisource.org/wiki/The_Foundation_of_the_Generalised_Theory_of_Relativity for a good translation. Accessed July 2016.

45 F.W. Dyson, A.S. Eddington, C. Davidson, "A Determination of the Deflection of Light by the Sun's Gravitational Field, from Observations Made at the Total Eclipse of 29 May 1919," Philosophical Transactions of the Royal Society 220A: 291–333, 1920.

2 The Big Bang proposal has a light travel issue as well. The cosmic background radiation is a faint glow of very low temperature radiation which is virtually the same in every direction. For the Big Bang cosmology, there are hot and cold parts of the early universe that have not had enough time to interchange radiation — but evidently they have done so. Thomas explains, "Hot and cold spots that lie on opposite sides of the visible universe are simply too far apart to have reached this same temperature even after 13.8 billion years."[46] This is called "the horizon problem" in secular astronomy and shows that a light travel problem occurs in classical secular cosmologies.

3 Many spiral galaxies appear the same age with distinct spirals no matter where they are found in deep space or nearby.[47] Again, as stated by Thomas: "Very distant spiral galaxies — where stars are arranged in great, winding arms — appear to have undergone the same amount of spiral arm winding as closer ones. This is consistent with the idea that astronomical time runs, or used to run, at very different rates than Earth time."[48]

4 It is of note that Scripture informs us that the Lord has "stretched out" the Heavens (Isaiah 40:22, 42:5; Job 9:8; Jeremiah 10:12; and Psalm 104:2). This suggests that there may have been an ordered expansion of space itself at the end of Day 4 of creation when he made the stars. This supports the proposal of Humphreys and Hartnett.

5 The Big Bang Theory relies on dark mass and dark energy to enable the proposal to work. Some dark mass has been observed but dark energy has not, and yet 75% of the universe is meant to be composed of this![49]

46 B. Thomas, "Distant Starlight and the Big Bang," *Acts & Facts*. 44 (6), June 2015.

47 A. Rigg, "Young Galaxies Too Old for the Big Bang," *Creation* 26(3), June–August 2004.

48 B. Thomas, "Distant Galaxies Look Too Mature for Big Bang," *Evidence for Creation*, November 2011.

49 J.A. Frieman, M.S. Turner, and D. Huterer, "Dark Energy and the Accelerating Universe," *Annual Review of Astronomy and Astrophysics*. 46: 386, 2008.

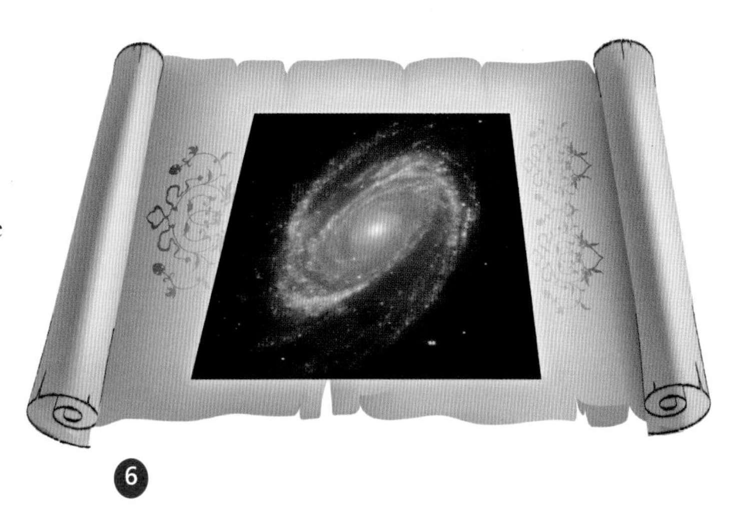

6

6 Isaiah 34:4 "The heavens shall be rolled up like a scroll."

6 To get the Big Bang model working, scientists have also had to rely on an inflation period in the first 10^{-36} seconds after the Big Bang event; in this very short time everything is supposedly growing at an exceedingly fast rate. However, this is a very speculative idea and is really a mathematical artifact to save the Big Bang model.

These issues of starlight and its reaching Earth from vast distances from Earth are further discussed on pages 127–128.

None can deny the magnificence of the night sky. The wonder of the stars as they roll across the heavens has been an inspiration for countless writers. Perhaps the most staggering statement in Scripture is that the Lord will roll them all up (figure 6) — this is stated in Isaiah 34:4, Psalm 102:25–26, and Hebrews 1:11–12. The verses in Hebrews 1 state "They will perish, but You remain; And they will all grow old like a garment; Like a cloak You will fold them up." This reinforces the understanding that space itself (made on the second day, according to Genesis 1) and expanded (see page 128 option 3), will one day collapse at His command. What a mighty God is reflected in the wonder of the heavens!

07 Starlight and Time

A common question people ask about creation in six days is how distant starlight could reach Earth in a very short amount of time.

1 Around 5,000 stars and a few galaxies can be seen from Earth with the naked eye.

Chapter 13 on pages 203–205 explains why the opening chapter of Genesis should be understood as describing six literal 24-hour days of creation.[1] This means that Earth and the stars cannot be billions of years old as most secular science claims. However, this raises a serious objection by some because many of the stars we see in the night sky are millions of light years away (figures 1 and 2). This means that it would normally take millions of years for that light to travel from its source to Earth. In fact, with telescopes it is possible to see stars and galaxies that appear to be billions of light years away.

The vast size of the universe and the enormous time normally required for light to travel across it, raises the question: "How could starlight travel billions of light years to Earth when Earth is only in the order of thousands of years old?" In fact,

1 For more detail on the six days see A. McIntosh, *Genesis 1–11: A Verse by Verse Commentary* (Day One Publications, Leominster. UK 2016). See also, Ken Ham, *Six Days* (Green Forest, AR: Master Books, 2013), and T. Mortenson, ed., *Coming to Grips with Genesis* (Green Forest, AR: Master Books 2008).

the starlight question should actually be: "How could starlight travel from distant stars to Earth in one or two days?" The reason for this is that it is reasonable to assume that Adam and Eve would have seen a similar night sky on Day 6 of creation that we see today.

Another specific starlight question concerns the timing of events we observe in the night sky. When viewing distant stars and galaxies millions of light years away, it can be observed that those bodies are undergoing movements such as receding or spinning. Therefore, there is not just the question of how distant starlight could travel so far in a short time but the additional question: "When did those events actually occur?" Before describing some of the main starlight and time theories in detail it is important to clarify that there are limits to what can be explained in the creation week.

The limits to explanations

When dealing with activities of the creation week it is not necessarily possible to explain "how" God created details like starlight because it is not usually possible to explain a supernatural act in terms of physical processes. Genesis teaches that God created a mature, fully-functioning Earth. We know that He created mature, fully-functioning plants, trees and fruit because Adam could pick fruit at the beginning of creation. We know also that Adam and Eve were fully grown and mature because they could speak and reason.

The principle of mature creation can also be seen in the miracles of Jesus. When Jesus turned water into mature, good quality wine (John 2), He did something supernatural that cannot be explained by natural processes. Likewise, when Jesus produced mature bread to feed thousands (Matthew 14, 15), He did something completely supernatural.

If starlight was brought to Earth supernaturally to create mature, fully-functioning starlight, then this places limits on what can be explained. So while it is legitimate to explore the question of how God may have caused starlight to reach Earth on Day 4 of creation, we must be open to the possibility that it was a supernatural act that cannot be explained by natural processes: "He does great things we cannot comprehend" (Job 37:5).

2 The Andromeda Galaxy can just be seen with the naked eye.

Young Earth starlight explanations

There are several possible explanations to the starlight question. Four are given here.

1. Light created in transit theory

When God created stars He may also have created the beams of light between the stars and Earth so that light was instantly observable from Earth. While this is a simple solution, it suffers from the problem that it would mean the events we observe in distant stars have not actually happened but are simply in the form of information in the beams of light. Most creation scholars today think it would not be God's way to create an appearance of events that never took place.

2. Speed of light decay theory

Some have proposed that the speed of light was infinite at the time of creation and has been decaying in accordance with a natural law of decay.[2] This would explain why starlight reached Earth by Day 6 and also why the speed of light is finite today. One of the reasons for proposing this theory is that some historical data indicates that the speed of light had been decaying in the past. However, further research of the evidence has shown that there is no clear evidence that the speed of light has been decaying. So this position is not generally accepted.

3. Relativistic time-dilation theories

Others have proposed that God used relativistic time dilation in such a way that vast amounts of time occurred in the distant universe while only a few days occurred on Earth. The concept involves the universe expanding rapidly during the creation week with Earth being near the center of the universe. This is discussed on page 125. Biblical support for this concept comes from verses that describe the heavens being supernaturally stretched out (for example: Isaiah 40:22, 42:5, 44:24). The fact that the Bible speaks of the universe being supernaturally folded up quickly at the end of time (Isaiah 34:4) also supports the idea that the universe was initially expanded outward rapidly.

These relativistic time dilation explanations involve extremely complex concepts from modern physics such as white holes, event horizons, and relativity. Essentially, these theories propose that there were very different time zones in the universe that allowed the creation of a fully-functioning universe on Day 4. There are two main versions of this explanation. The first was put forward by Russell Humpreys[3] and the second by John Hartnett.[4] Relativistic time dilation involves a mix of supernatural and natural processes (stretching out) and has the advantage that it has a degree of predictive power in the way it can explain some of the features of the universe we observe today.

4. Supernatural speeding-up of starlight

A recent and more straightforward explanation involves God supernaturally causing a fast propagation of light on Day 4 of creation by stretching the space of the universe. As with time dilation theories, biblical support comes from the many verses that refer to the heavens being "stretched out."

The supernatural speeding up of light by the stretching of space was proposed by this author in 2002.[5] However, it has also been explained more recently and in more detail by Danny Faulkner, who calls it the Dasha' solution. (*Dasha'* comes from the Hebrew for "bringing forth.")[6] In his book, Faulkner argues that starlight was supernaturally speeded up in the same way that plants were supernaturally and rapidly grown from seeds on Day 3. The fact that Scripture describes plants as "brought forth" from the Earth supports the idea that plants grew very quickly from seeds rather than being created as fully grown. If plants were supernaturally and rapidly grown, that would support the idea that starlight was supernaturally and rapidly brought to Earth in a very short amount of time. Faulkner uses the analogy of a time-lapse camera to imagine the fast growth of plants or starlight.[7] This author (Stuart Burgess) uses a similar analogy of a fast-forwarding video to imagine the supernatural speeding up of starlight.[8]

If God did supernaturally speed up starlight, then the events we see in the stars would all have actually happened, but they would have happened within the last few thousand years. Both Faulkner and Burgess have explained that this supernatural explanation is not so different from the relativistic time dilation explanation. However, it relies on God supernaturally speeding up light to bring it rapidly

2 B. Setterfield, "The Velocity of Light and the Age of the Universe," Part 1, *Ex Nihilo*, vol. 4, no. 1, 1981.

3 D.R. Humphreys, *Starlight and Time* (Green Forest, AR: Master Books. 1994).

4 J. Hartnett, 2003. "A New Cosmology: Solution to the Starlight Travel Time Problem," *TJ* 17, no. 2:98–102.

5 S.C. Burgess, *He Made the Stars Also* (Leominister UK: Day One Publications, 2002).

6 D. Faulkner, *The Created Cosmos* (Green Forest, AR: Master Books, 2016), p. 216.

7 D. Faulkner, "A Proposal for a New Solution to the Light Travel Time Problem," *Answers in Genesis Journal*, July 24, 2013; last featured February 23. 2014.

8 S.C. Burgess, *He Made the Stars Also* (Leominister UK: Day One Publications, 2002).

3 Stars in the Orion constellation are approximately 300 to 1,400 light years away but would probably have been seen by Adam and Eve. *(Photo from Wikimedia)*

to Earth on Day 4 (figure 3) and it does not rely on vast periods of time occurring in deep space.

When the Bible states, "For he spoke and it was done" (Psalm 33:9), this may well refer not just to the creation of stars, but also to the bringing of starlight to Earth, supporting the theory that starlight was supernaturally speeded up. An advantage of this theory is that it does not rely on complex mathematical and physical concepts!

Summary

There are several explanations of how God may have caused starlight to reach Earth in a short amount of time. The authors of this book have a preference for the third and fourth models presented in this chapter. One thing is sure, God is all-powerful and well capable of supernaturally creating mature starlight whether we can explain it or not. While the question of starlight and time is difficult to explain, it is a very minor challenge compared to the major deficiencies of the Big Bang theory.[9]

9 D. Faulkner, *Big Bang — The Evolution of a Theory*, *Answers Magazine*, AIG, October 1, 2013.

08 Beauty — Flowers

The great variety of colors, shapes, and scents of flowers makes them among the most beautiful parts of creation.

1 Bluebells are suited to growing in woodlands like these in Nyman's Garden, Sussex, UK. *(Photo Brian Edwards)*
2 A field of buttercups on the island of Iona off the west coast of Scotland. *(Photo: Brian Edwards)*

Even though beauty is subjective, there are nevertheless objective reasons why something can be considered beautiful. Visual beauty is created through features like geometric shapes, blending, borders, and bright color. A surface is beautiful to touch through features like smoothness and softness. Flowers have an abundance of such properties.

A beautiful fragrance is created by the emission of pleasant-smelling volatile odor molecules. The fragrance from a single flower may contain up to 150 different types of odor molecules that combine to give a unique and beautiful smell. The quality of flower fragrance is such that some flowers are used to produce very expensive perfumes.

Bluebells

Bluebells are loved for their striking purple-blue color and bell shape. The petals at the end of the stem of the flower join together to produce the delicate and beautiful hanging bluebells. Bluebells are able to live in woodland because they are shade-tolerant and because they bloom in springtime just before leaves block out too much light. One of the reasons why a wood carpeted with bluebells (figure 1) is a lovely spectacle is because it is a reversal of the color scheme in open countryside of green ground and blue sky. The blue pigment of bluebells comes from a very complex chemical compound called *malonylawobanin* which includes the chemical components *delphinidin, malonic acid, and coumaric acid.*

Buttercups

Buttercups are so called because they have a bright glossy yellow color and because the five petals are curled and overlapping to give a cup shape (figure 2). The yellow color of buttercups comes from a carotenoid pigment in the thin outer layer of the petals called the epidermal layer which absorbs light in the blue and green region of the optical spectrum. The result is that mostly yellow is reflected. The

(Photo: Brian Edwards)

3 Violet aster flowers
4 A beautiful lily with an exquisite scent *(Lilium Trumpet Regale)*

reason for the high gloss of the buttercup is that the epidermal layer has two extremely flat surfaces, separated by a gap of air. Reflection of light by the two smooth surfaces and by the air layer produces a high degree of gloss.[1] The design of the buttercup has amazed scientists because of its precision geometry and effectiveness in producing a glossy effect.

Aster flowers

Aster is the Latin word for "star" and describes the starry layout of flower heads on aster flowers like the daisy, the sunflower, and the violet aster, as shown in figure 3. Asters are actually composite flowers consisting of many flowers that, grouped together, look like one flower. The yellow central part of the daisy flower head looks as though it is full of yellow stamens, but it actually consists of many yellow disc flowers. Around the group of yellow discs are white ray flowers that form a complete disc or "whorl." The perfect coordination of hundreds (sometimes thousands) of individual flowers to form what appears to be a single flower is truly remarkable and bears testimony to the great wisdom of the Creator.

Cultivated flowers

Of course, many flowers we enjoy in gardens, like lilies (figure 4) and roses (figure 5), have been cultivated by gardeners over hundreds of years to emphasize certain beautiful features. Such cultivated flowers still show God's wisdom because God created the genetic potential there in the first place to produce the designs we see today.

5 A cultivated rose

Roses have several nested "whorls" of petals that combine to produce the lovely rose effect. Lilies are large, strongly scented flowers with a range of colors, including whites, yellows, oranges, pinks, and reds with markings like spots and brushstrokes. Lilies have 3 petals and 3 sepals that look the same, and in this case the petals and sepals are both called "tepals."

Why are flowers beautiful?

The beauty of flowers is far beyond what is needed to attract insects. Jesus said that God had clothed flowers with a glorious splendor that exceeded the splendor of King Solomon (Matthew 6:29). The clear implication is that God created beauty for the sake of beauty and not just as a means for insects to be attracted to flowers.

1 Silvia Vignolini et al, "Directional Scattering from the Glossy Flower of *Ranunculus*: How the Buttercup Lights up Your Chin," *Journal of the Royal Society Interface*, December 14, 2011.

Trees

Trees are beautiful because of their great size, yet they also contain delicate features such as leaves, blossoms, fruit, and even sweet smells.

1 Jacaranda trees in blossom

Trees produce many important things for life on Earth, including oxygen for Earth's atmosphere, a wide range of fruits and nuts, a diverse range of woods, and an extensive habitat for many creatures. However, another important function of trees is to help make creation beautiful. Genesis 2:9 teaches that "God made every tree grow that is pleasant to the sight" (NKJV), so the beauty of trees should be seen as a deliberate design feature.

Beautiful coloring

Springtime blossoms can give a tree a delicate floral attractiveness. Some of the most well-known blossoms are the pink blossom of Japanese cherry trees and the lilac-blue blossom of the Jacaranda tree which is native to South America (figure 1).

Even though most tree leaves are colored green, there are over a million shades of green that can be detected by the human eye. Some leaves have large quantities of red pigments as well as green chlorophyll that give them a copper color, as with the copper beech tree. In autumn, the leaves of many trees turn into beautiful colors of yellow, orange, and red. The yellow, orange, and red pigments are always in the leaves but they are swamped by the green chlorophyll while the leaves are alive. In the autumn the trees break down the chlorophyll and nutrients and the other colors remain. Many

2 Leaf designs from some common trees
3 A mature sequoia tree

scientists are mystified as to why leaves should contain colorful pigments because they are unnecessary for the tree's survival. However, such coloring is what would be expected from a Creator.

Beautiful variety

There are over 23,000 species of trees in a vast range of sizes and designs. Softwood trees, like fir and pine, are beautiful for their tall straight trunks, evergreen color, and intricate cones. Hardwood trees, like ash, birch, beech, oak, elm, and cherry, are beautiful for their variety of shapes and broad leaves. Leaves come in a myriad of sizes and shapes (figure 2) and even the needles of conifers have a variety of designs. Some trees have coarsely grained bark, like willow, while others have smooth bark, like beech trees.

There is even a variation in the grain structure of different woods. Some woods, like teak and rosewood, have beautiful grains when cut and polished, which makes them ideal for furniture. Others have pleasant aromatic scents, such as the lemon eucalyptus, cedar, and sweet birch. It is estimated that there are over three trillion trees on Earth, yet every one is unique.

Beauty through size

The towering size of mature trees is beautiful because of the way such trees provide shade and because they can be viewed from far away. Many trees, like oaks, grow to over 131 feet in height, and some, like the California redwood, grow to over 328 feet. A single sequoia tree (figure 3) can be up to 98.5 feet in circumference and can hold up to 2 billion leaves. Sometimes it is the vast quantity of fruit that creates a beautiful effect. A single apple tree can produce up to 1,000 apples each year and a single nutmeg tree up to 10,000 nuts.

Beautiful music

There are particular types of wood that make high quality musical instruments. These are called "tone woods" because they produce the pure resonance needed in stringed instruments. Spruces are often used for the sound boards of stringed instruments like violins and guitars, because of the high stiffness-to-weight ratio. Sound boards are the layer of wood that is just under the strings and are the main reason for the production of sound. Maple is often used for the backs and sides of woodwind instruments. There is no doubt that God thought of man's needs in areas like music, furniture making, and house construction when He created trees.

The Color Scheme of Creation

We often take the color scheme of creation for granted, but when we consider the details we can see it is supremely well coordinated with the right contrasts and background colors.

1 The blue-green contrast of sky and land

Green plants

Green is the dominating color of plants such as trees, shrubs, bushes, and grass because green chlorophyll is the pigment in the leaves which converts sunlight into chemical energy. Green is also the ideal dominating color because it is the most restful and relaxing color for humans. In contrast, if we were surrounded by large amounts of red it would raise our blood pressure and not be restful. It is interesting to note that human eyes have more photoreceptor cells for the color green than for any other color, indicating that God has designed humans to appreciate the restfulness of green plants.

Another important design feature of green is that it is just the right background color for bright flowers. One of the reasons for the beauty of a garden is not just the bright yellows, reds, and blues of the flowers, but the fact that green is the best background for appreciating bright colors.

Contrasting land and sky

One of the reasons for the beauty of landscape is the contrast between the blue sky and green land (figure 1). How strange it would be if both the Earth and the sky were green or both blue! White clouds also makes a good contrast with the land. A blue/white sky contrasts not just with green plants but other common land surfaces of desert sand, rock, and mud.

Such a color scheme is not the result of chance but design and planning. When God designed chlorophyll for plants, he deliberately made it a green color in order to produce the right dominating color for the land. And when God designed air molecules, he deliberately made them scatter light so that the sky would be blue.

Another beautiful feature of the color scheme in creation is that blue is the most uncommon color for plants and trees. This produces the most beautiful effect because when blue dominates in the sky it is best if it is the least common color on the ground.

Seasons

Even though seasons probably existed only after the worldwide Flood (Genesis 6), they are nevertheless wonderfully designed. Many places around the world experience beautiful changing landscapes throughout the year. Spring has the beauty of vibrant greens and bright blossoms. Summer offers colorful flowers and exquisite insects like butterflies, while autumn brings yellow, orange, and red leaves (figure 2), and even winter has its own beauty with white frosts and snow (figure 3). On one day there can be beautiful changes throughout the daylight hours, starting with a stunning sunrise and finishing with a beautiful sunset. Even though air is transparent, God has skillfully designed it to be capable of making a blue sky or a red sunrise or sunset.

Beautiful night lights

Earth is positioned about two-thirds from the center of the Milky Way Galaxy, which is ideal for an awesome night sky. If Earth was near the outside of the galaxy, it would be extremely dark at night; by contrast, if Earth was near the center, the night would be nearly as bright as the day. But the precise position of two-thirds from the center is ideal because it means that a few thousand stars can be viewed with the naked eye; this is enough to be fascinating but not so many as to be too bright. The same perfect spacing can be seen in the way galaxies are distributed. Other galaxies are not too close to make the nighttime sky too bright, but they are not so far away that they cannot be seen with telescopes. Even the design of the moon is ideal because it is not always in the nighttime sky and therefore it does not always dominate the night sky.

2 Autumn scene from Nagoya, Japan
3 Winter scene

Planned and made

When you consider the color scheme of creation, it is as if an expert in art has coordinated and planned the colors with great care and attention. The Bible reminds us that God is perfect in knowledge (Job 37:14), so we should not be surprised at such a beautiful color scheme.

Bird Plumage

Many birds have beautiful colors and patterns in their plumage. In many cases, the beauty cannot be explained by survival needs but represents a deliberate "added beauty" put in place by the Creator.

1 Flock of flamingos

The colors and patterns in bird feathers help birds to recognize their own kind. However, many feathers are far more beautiful than they need to be. This purposeful overdesign can be termed *added beauty* or beauty for beauty's sake.

Colors in feathers are produced in three ways: pigments, optical effects, or a combination of both. For example, the browns and blacks of many common birds (like sparrows and crows) are produced by the melanin pigment. In the case of flamingos, they have pink feathers (figure 1) because of the carotenoid pigment which they obtain in their diet.

There are several remarkable optical effects that produce colors in birds. Peacocks produce bright colors through thin-film interference where light reflects off of very thin films of keratin. The white in swan feathers is produced when light is reflected and refracted within the transparent structure of the feathers. The blue in budgerigar feathers is produced when tiny air-filled cavities scatter light (preferentially scattering blue because of its higher frequency). Some feathers have colors which are a combination of pigments and structural colors as with many green feathers.

The rest of this chapter will focus on how thin-film interference is produced in peacock feathers. It should be noted that thin-film interference is found in many other birds such as pigeons, hummingbirds, and birds of paradise.

Peacock display feathers

Peacocks are male peafowls and are native to parts of Asia and Africa. The peacock is considered to be so special that it is kept in zoos and gardens around the world. It is not just the tail feathers that contain striking beauty. The peacock also has a beautiful blue neck, crest feathers on the head and white borders to the eyes (figure 2). The deep blue colors and crest feathers create a majestic royal style.

An adult peacock has around 200 tail feathers that can be deployed into a very impressive fan formation (figure 2). An incredible feature of this is that it forms an angle of nearly 360 degrees in order to surround the entire body. In addition, the axis of every feather projects back to an approximately common geometric center. The precise radial alignment of feathers requires the root of each feather to be pointed with a high degree of accuracy. Another amazing feature of the displayed feathers is that they are deployed and held in position by muscles in the tail. A particularly beautiful feature of the fan formation is the uniform spacing of the eyes (figure 2). Even though the display contains around 170 eye feathers, all of the "eyes" are visible and are spaced evenly apart. Each of the eyes is

2 Peacock with tail feathers deployed

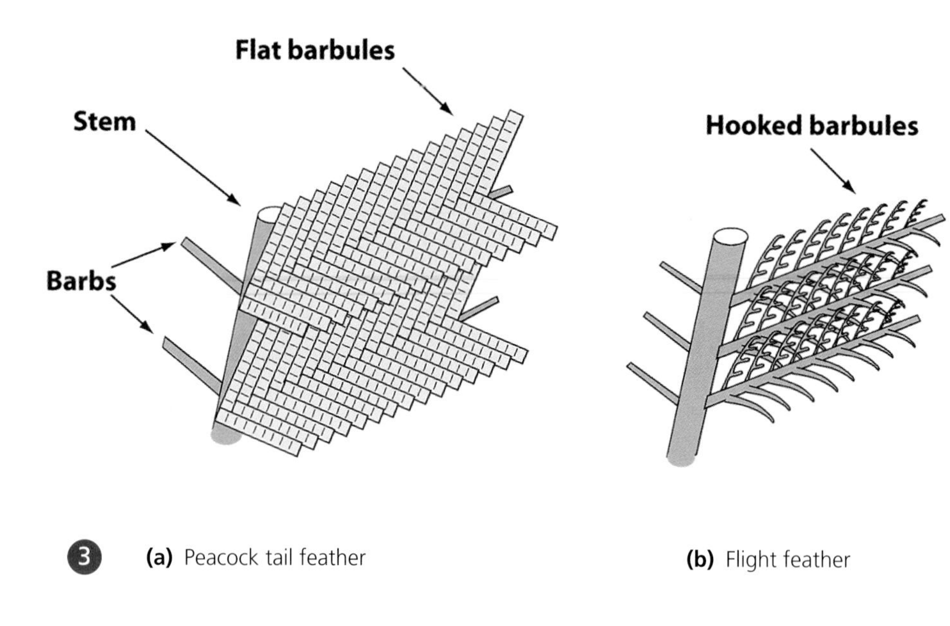

3 Comparison between a peacock tail feather and a flight feather

(a) Peacock tail feather

(b) Flight feather

visible because the feathers are layered with the short feathers at the front and the longer feathers at the back and with each feather having just the right length to produce even spacing.

Structural coloring

Most colors we see around us are due to pigments such as the green chlorophyll in plants. However, the colors in peacock feathers are mostly due to thin-film interference. Peacock tail feathers have a hierarchical structure with barbs coming off the main stem and barbules coming off the barbs (figure 3). Each tiny barbule contains a number of minute segments (just visible in figure 3), each of which can make a different color.

Each segment has layers of keratin that are so thin that they are comparable with the wavelengths of colored light. When white light is reflected by the layers, some colors are removed and the reflected light has a strong color (figure 4). Because the colors are due to an optical effect, they do not fade and have a deep luster. The colors even change with the angle of view, producing an effect called iridescence (figure 4).

The "eye" feather

The "eye" in the eye feather is formed from a digital pattern produced by the precise coordination of many thousands of individual segments (figure 5). Amazingly, the different colors are due to precise differences in the design of thin-film layers across the feather. The "pupil" of the eye is formed by a dark purple cardioid and the "iris" is formed by a blue ellipsoid. These shapes are located within a pointed bronze ellipsoid that is surrounded by one or two green fringes.

As well as the beauty of the colors in the "eye" feathers there are several other tiny details of design that often go unnoticed. The absence of the stem in the upper half of the eye pattern (due to clever changes in the angle of the barbs as shown in figure 5) is an important feature, because it greatly reduces the disruption of the stem on the pattern. The fact that the stem is narrow and colored brown also helps the pattern not to be spoiled (figure 5). There is even contrast between loose barbs and tight barbs around the edges of the feather (figure 5). It is important to realize that each of these tiny features is precisely and deliberately specified in the DNA.

Added beauty

All the beautiful features of the peacock, like the thin-film interference, the precise spacing of the eye feathers, the crest feathers, and the detailed features of the eye feather, require huge amounts of precise genetic information. There are sexual selection

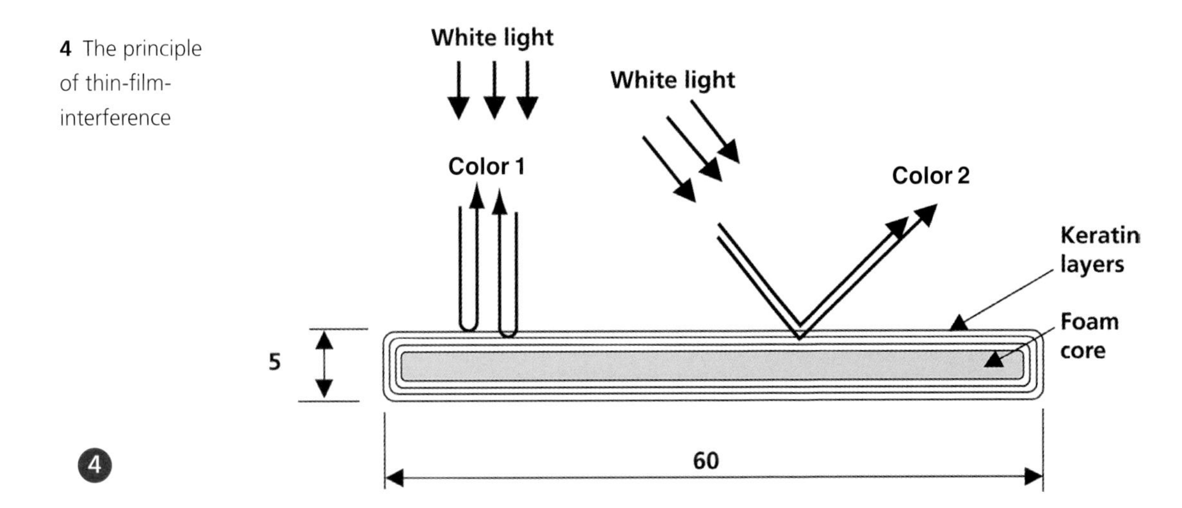

4 The principle of thin-film-interference

theories that claim that female peafowls prefer beautiful males and that this explains the existence of beauty. However, there are major problems with such theories.[2] Much of the intricate optical design involves irreducible complexity and so cannot evolve step by step. And even if it could evolve in steps, there is no reason why a peafowl should select beauty rather than random or ugly features. For example, why should peahens prefer the eye pattern to be evenly spaced in the fan display? And why should they prefer there not to be a stem in the upper half of the eye pattern?

Field studies have shown that the beauty of birds like peacocks is actually degrading over time, not improving.[3] This confirms the predictions of creationists that creation is degrading not evolving. (See pages 54–55 for more on the feathers of peacocks.)

2 S. Burgess, "The Beauty of the Peacock Tail and the Problems with the Theory of Sexual Selection," *Journal of Creation* 15, no 2 (August 2001): 94–102.

3 J.J. Wiens, "Widespread Loss of Sexually Selected Traits: How the Peacock Lost Its Spots," *Trends in Ecology and Evolution* Volume 16, Issue 9, 1 September 2001, p. 517–5235.

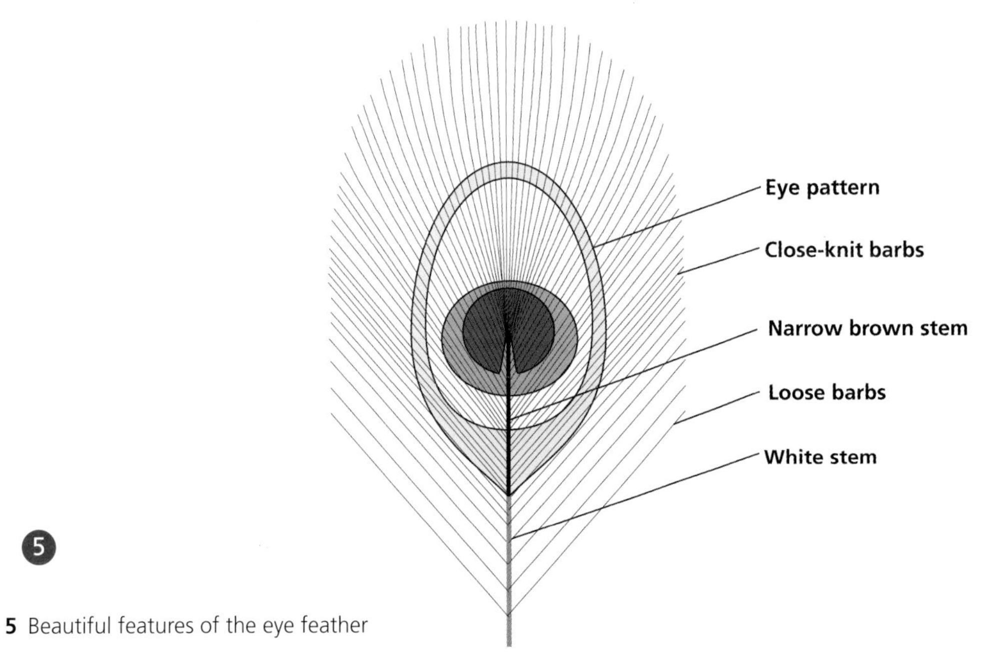

5 Beautiful features of the eye feather

Birdsong

Songbirds are tiny creatures, yet they have powerful voices that can fill the air with melodic music that contains many features of advanced musical beauty.

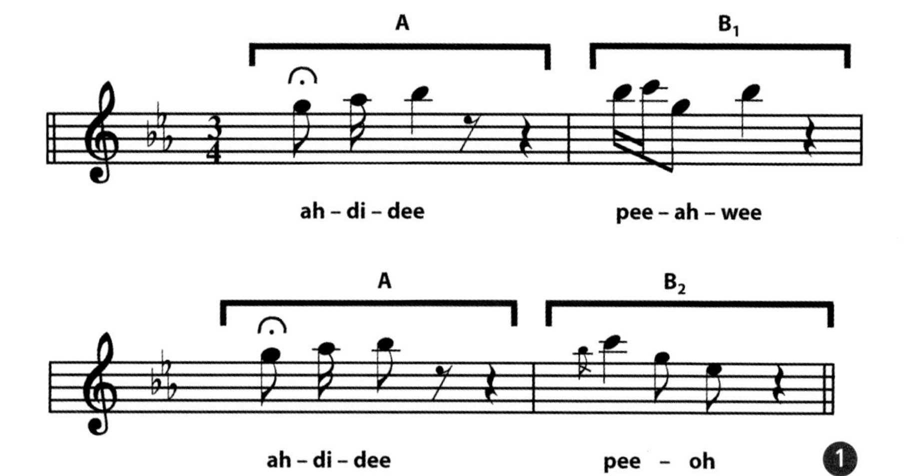

1 The song of a wood pewee

2 A wood pewee singing

Of the 9,000 species of bird in the world, around 4,000 are songbirds. Some of the most common songbirds in Europe include robins, blackbirds, thrushes, chaffinches, warblers, and nightingales. For more on birdsong and the syrinx which enables this, see pages 58–59.

Melodic songs

Songbirds do not sing random notes but specific songs that have musical structure such as a time signature, key signature, and phrases. If you listen carefully to a songbird over several days, you will notice that it has a repertoire of songs and that these are repeated with remarkable accuracy. The nightingale holds the record for the largest repertoire with up to 300 songs. Some birds, like marsh warblers, have an ability to copy the songs of other species.

A typical birdsong lasts between 4 and 20 seconds and contains a few hundred notes that are structured into separate musical phrases. Musical structure does not happen by chance, because to compose a tune with a key signature and time signature requires the selection of very particular notes with precise timing.

Figure 1 illustrates the song of the North American wood pewee (figure 2) This section of the song consists of four phrases and contains melody, rhythm, variety, and unity. Of the four phrases shown, the first phrase is repeated almost exactly in the third section and the other two phrases are an appropriate and pleasant variation on the theme. The song ends with an interval, called a major third, that brings the section of music to a clear conclusion. Composers are taught that there are only a few intervals that can conclude a song

Key: X = 1st bird, Y = 2nd bird

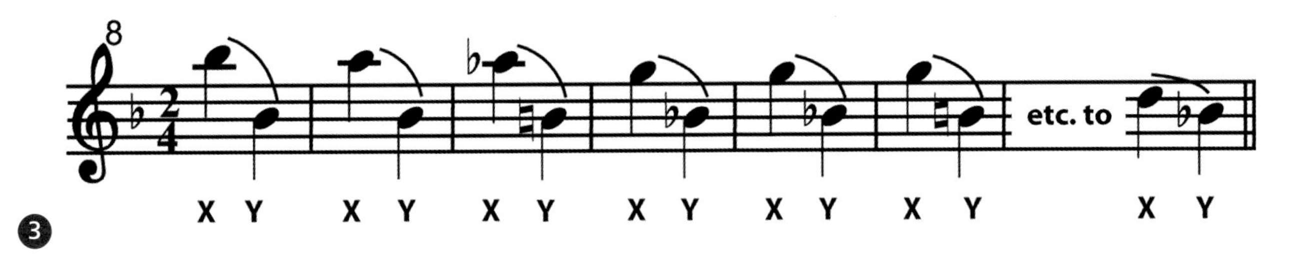

3 A duet sung by a pair of African shrikes **4** A Blue-throated robin enjoying a song

with "finality," such as a major third or major fifth, yet many birds finish their songs with these very precise intervals.

Counter-singing, duets, and quartets

Some songbirds perform matched counter-singing where two neighboring birds take turns singing and reply with an appropriate melody. This is a very common occurrence in the summer months with male blackbirds.

Some mating pairs of songbirds, like African shrikes (figure 3) and African robins, (figure 4) can perform what is called antiphonal singing, where two birds sing alternative notes in a duet. To sing such duets requires that both singers know the song exactly and sing at the same pitch and speed. An example of a duet from the African shrike is shown in figure 3. African shrikes can even sing in quartets. In this case, there is a matched duet where one pair of birds sings a duet and then a matched duet is sung by another pair of birds.

The origin of birdsong

It is mainly male birds that sing, and they sing throughout the year to mark out their territory. However, when marking out territory there is no advantage in songs being beautiful. Many birds advertise their territory with extremely simple and crude sounds. Evolutionary theories have to put forward bizarre theories such as beautiful songs being more frightening to neighbouring birds and therefore songs become more beautiful over time as they become more frightening.[4] But, as with so many evolutionary theories, this particular theory is full of holes and contradictions.

Human composers require many years of training to compose music with pleasant-sounding structure and melody. Yet birds have no training at music school! The only explanation for the beauty of birdsong is that it was created by God for the sake of creating beauty. Recent research has shown that humans and songbirds have the same singing genes that enable the creation and production of music.[5] This provides wonderful evidence of common design by a common Designer.

Belief in creation increases a person's appreciation of beauty because they know beauty is the work of a loving Creator.

4 C.K. Catchpole and P.J.B. Slater, *Bird Song: Biological Themes and Variations* (Cambridge University Press) 1995.

5 Colin Barras, "Humans and Birds Share the Same Singing Genes," *New Scientist*, December 11, 2014.

09 **Mathematics and Beauty** — The Nature of Mathematics

Even the logical world of mathematics displays design and beauty.

1 Where does the very concept of number and rationality come from?

First of all, we should understand what mathematics is. It is an expression of rational thought and reasoning. It is the language of logic and precision, and as such only makes sense because reason and logic exist in the first place. The very fact that we can think logically is a demonstration that we are made in the image of God. Although animals like dogs and dolphins can be trained to do tricks and another generation following them can do the same tricks, what they cannot do is pass on technical instructions (verbal and certainly not written) to the next generation such that it builds on the previous generation's discoveries. Logical constructs are very much part of this (figure 1) and mathematics is vital for expression of technology. The concept of number is especially fundamental to this discipline.

The ability of humans to think abstractly in the arts and the sciences is unique. In both areas there is structure of thought which involves a representation of that idea / thought / concept / number. Such an idea / thought / concept / number is real, but in itself it is not physical. It may lead to a painting, architecture, or a machine of some kind, but it does not have to. The thought may be about something physical, but the reality of the thought is different and exists in its own right. Concerning mathematics, Lisle has rightly stated, "Numbers are concepts. Thus they are abstract in nature. They exist in the world of thought and are not material or physical. You cannot literally touch a number, or even see one, because they are not made of matter."[1]

We can see a representation of number (the language of mathematics), but the number itself has an existence without that representation. In scientific analysis the representation of logical engineering principles involves mathematics (see for instance figure 2) which can be handed down across generations. As C.S. Lewis once stated in his brilliant book *Miracles*:

1 J. Lisle, "Evolutionary Math?" *ICR Acts & Facts* 41 (12) 11–13, 2012.

2 One of the biggest hurdles for understanding is the concept of fractions!

> Nature is quite powerless to produce rational thought: not that she never modifies our thinking but that the moment she does so, it ceases (for that very reason) to be rational. For, as we have seen, a train of thought loses all rational credentials as soon as it can be shown to be wholly the result of non-rational causes.[2]

If it could be demonstrated that our ability to reason that 2 plus 2 equals 4 was a result of some chemistry or other principle of energy transfer/ neuron transmission in our brains, then the rational basis for logic would be lost! Rationality has to be transcendent to matter to make any sense at all. This is why atheism is in essence cutting off the very branch of reason that it is sitting on! Leading thinkers who are not necessarily believers have realized this — notably in the last century the British born geneticist John Haldane[3] and today the U.S. philosopher Thomas Nagel.[4] Famously, Einstein stated, "The most incomprehensible thing about the universe is that it is comprehensible"[5] And Nagel said, "Evolutionary naturalism provides an account of our capacities that undermines their reliability, and in doing so undermines itself.'[6]

The rationality of the universe cannot be from the universe itself, and coupled with this, is a strange beauty which speaks of the mind behind it all. Mathematics itself speaks of the Lord who made our brains to think.

2 C.S. Lewis, *Miracles* (Harper Collins), p. 38–39, 1947 (reprint 2012).

3 J.B.S. Haldane, *Possible Worlds* (1927). See p. 209 where he succinctly states the issue "If my mental processes are determined solely by the motions of atoms in my brain, I have no reason to suppose that my beliefs are true ... and hence I have no reason for supposing my brain to be composed of atoms."

4 T. Nagel, *Mind and Cosmos — Why the Materialist Neo-Darwinian Conception of Nature Is Almost Certainly False* (New York: Oxford University Press, 2012). See p. 15–16, "Materialist naturalism leads to reductionist ambitions because it seems unacceptable to deny the reality of all those familiar things that are not at first glance physical. But if no plausible reduction is available, and if denying reality to the mental continues to be unacceptable, that suggests that the original premise, materialist naturalism, is false, and not just around the edges. Perhaps the natural order is not exclusively physical."

5 A. Einstein, "Physics and Reality"(1936), in *Ideas and Opinions*, Sonja Bargmann, trans. (New York: Bonanza, 1954), p. 292.

6 T. Nagel, *Mind and Cosmos* (Oxford University Press, 2012), p. 27.

Patterns and Beauty in Mathematics

There is a peculiarity of mathematics such that the very language of mathematics is connected with patterns and has an aesthetic appeal.

$$
\begin{aligned}
1 \times 9 + 2 &= 11\\
12 \times 9 + 3 &= 111\\
123 \times 9 + 4 &= 1111\\
1234 \times 9 + 5 &= 11111\\
12345 \times 9 + 6 &= 111111\\
123456 \times 9 + 7 &= 1111111\\
1234567 \times 9 + 8 &= 11111111\\
12345678 \times 9 + 9 &= 111111111\\
123456789 \times 9 + 10 &= 1111111111
\end{aligned}
$$

1

$$
\begin{aligned}
1 \times 8 + 1 &= 9\\
12 \times 8 + 2 &= 98\\
123 \times 8 + 3 &= 987\\
1234 \times 8 + 4 &= 9876\\
12345 \times 8 + 5 &= 98765\\
123456 \times 8 + 6 &= 987654\\
1234567 \times 8 + 7 &= 9876543\\
12345678 \times 8 + 8 &= 98765432\\
123456789 \times 8 + 9 &= 987654321
\end{aligned}
$$

2

1 / 2 There are patterns and elegance in numbers themselves.

Numbers can be used with a different base number. In our decimal system, that base is 10. In computers, the fundamental base is 2, so that there are two binary states "on" or "off," 1 or 0. So counting from 0 to 5 in binary is written as 000, 001, 010, 011, 100, 101. Everything in computer codes will finally be transmitted to a location which is either on (1) or off (0). Patterns involving 1 and 0 will emerge across the sequence of coded instructions. This was literally the case in the early digital computers which were fed by paper tapes and cards with or without holes representing these instructions.

The interesting fact is that for the higher base that we use (10) significant patterns emerge. Examples of two of these are shown in figures 1 and 2. There are elegant patterns in the numbers themselves expressed this way which would not have been apparent to the Romans who used a different means of counting (V stood for our 5, X stood for our 10, L stood for our 50, so 8 was VIII and 9 was IX, 54 was LIV and so on). Though it was logical, it was not the most elegant way of writing numbers. So there is a principle that when mathematics is put succinctly, it often carries with it an elegance and a beauty. The arch atheist Bertrand Russell had to admit:

> Mathematics, rightly viewed, possesses not only truth, but supreme beauty — a beauty, cold and austere, like that of sculpture, without appeal to any part of our weaker nature, without the gorgeous trappings of painting or music, yet sublimely pure, and capable of a stern perfection such as only the greatest art can show. The true spirit of delight, the exaltation, the sense of being more than Man, which is the touchstone of the highest excellence, is to be found in mathematics as surely as in poetry.[7]

7 Bertrand Russell (1872–1970) *Mysticism and Logic and Other Essays* (London: George Allen and Unwin, 1959), chapter 4, "The study of Mathematics," p. 61.

The Fibonacci series and the Golden Ratio

There are sequences associated with growth that are found repeatedly in nature. These cause attractive spiral patterns to form and the actual count of them follows the Fibonacci series given by 0, 1, 1, 2, 3, 5, 8, 13, 21, 34, 55, 89, 144 . . . where the next number in the series is found by adding the previous two together. The ratio of two consecutive numbers tends to the value 1.618 which can be written more exactly as

$$\Phi = (1+\sqrt{5})/2$$

This number is called the golden ratio and was known long ago by the ancient Greeks, as it arises in the geometry of the regular pentagon, and is associated with what are pleasing proportions to the eye. Painters naturally will tend to offset objects away from the center by that ratio, and architecture such as the Pantheon in Greece uses this ratio.

This aspect of beauty is mentioned in chapters 4 and 8 (pages 54–55 and 136–139) concerning peacock feathers. The golden ratio also occurs in many other aspects of nature[8] — notably fir cones (figure 3), pineapples (figure 4), sunflowers (figure 5), and other plants that have seeds which go out in spirals (see figures 3/4). For these, the number of spirals follows this series. For pineapples, the number of spirals is usually the Fibonacci numbers 5, 8, and 13. For sunflowers (part of the daisy family), they usually have 55, 89, or 144 petals, and the number of left-handed and right-handed spiral patterns so evident in their seed pods is usually 21, 34, and 55. The direction of the spirals is linked to the angles 222.5° and its complement 137.5°, which

8 P. Lynch, "Sunflowers and Fibonacci: Models of efficiency," ThatsMaths blog June 5, 2014. See https://thatsmaths.com/2014/06/05/sunflowers-and-fibonacci-models-of-efficiency/. Accessed Sept 2016.

3

4

3/4 The Fibonacci spirals on fir cones and pineapples

5 Fibonacci spirals on sunflowers

6 Sunflower packing options for n =1,000 seeds using the formula $r(n) = \sqrt{n}$ and $\theta(n) = 2\pi n / \Phi$ where r,θ is the coordinates of the location (radius r and angle θ) where the next seed is placed. Φ=1.618 corresponds to the golden angle case, and plausible packing examples are demonstrated for two cases away from this exact value.

are linked to the golden ratio.[9] Further information is given in the work of Vogel[10] and Chandler.[11]

The patterns formed by Chandler's method are shown in figure 6. Though it is argued that this is just part of a natural solution to a packing problem for seeds, we show in that figure that there are other very tight packing solutions for *different* values of Φ but that in nature it is always 1.618 that is used. Both efficiency and beauty are thus demonstrated in nature.

Φ=1.1733

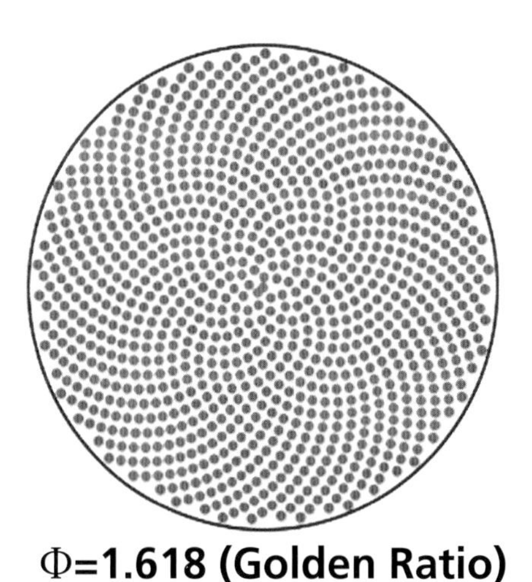

Φ=1.618 (Golden Ratio)

6

Φ=1.5856

9 Associated with the golden ratio Φ there is a golden angle 222.5° (and its complement 137.5°). Dividing a circle into two arcs with the ratio 1.618 leads to the longer arc subtending an angle of about 222.5° or, more precisely, $2\pi/\Phi$ radians. This and the complement are called the "golden angles." Helmut Vogel (ref. 10) devised a simple mathematical description for the geometry of sunflower seed patterns and a variation of that by Chandler (ref. 11) is used here given by $r(n) = \sqrt{n}$ and $\theta(n) = 2\pi\, n / \Phi$ where Φ is the golden ratio 1.618. Thus, as n increases by one, the position rotates through the larger golden angle and the radius increases as the square root of n.

10 H. Vogel, "A Better Way to Construct the Sunflower Head," *Mathematical Biosciences* 44 : 179–189, 1979.

11 D. Chandler, "The Golden Number," Turbomanage web site: https://turbomanage.wordpress.com/phi/. Accessed September 2016. The figures shown in figure 6 were made using his computer code. The author (Andy McIntosh) is grateful to David Chandler for his work and his willingness to have these figures used here from his computer code.

Mathematics and Music

There is a peculiarity of mathematics such that the very language of mathematics is connected with patterns and has an aesthetic appeal.

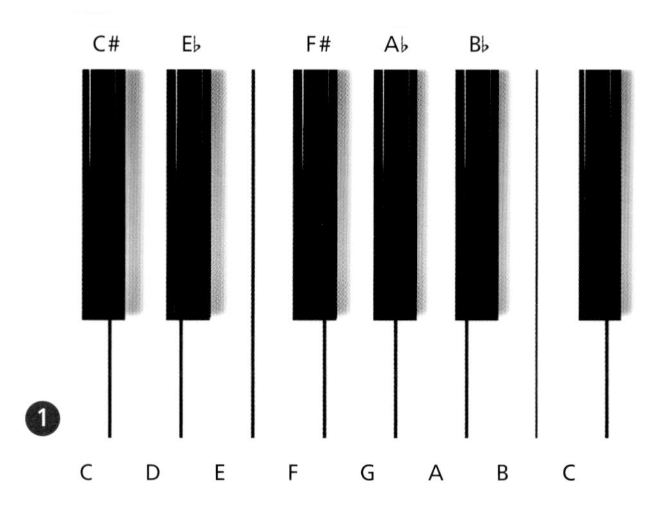

1 The 13-note western musical scale

It is hardly surprising that some well known mathematicians like Einstein have also been musicians,[12] since music is built on patterns which are aesthetically appealing.

Although it is true that there is a large subjective element in the appreciation of music, there is clearly much also that is to do with structure and patterns. It is particularly true in classical western music since in many respects this grew out of the order and worldview connected to the earlier Reformation across Europe.

The western musical scale (figure 1) used today has the octave divided into twelve equal parts, with each semitone (half-step) as an interval of exactly the twelfth root of two. Consequently, on a logarithmic scale (figure 2) each of these have the same frequency difference, and twelve of these equal half-steps add up exactly to an octave. This is called the equally tempered musical scale.[13] So all notes are related to a base frequency of A_4 (that is A in the 4th octave) which is 440 Hz, and any note has a frequency calculated by the formula $440 \times 2^{n/12}$ where n is the number of semitones in either direction from A_4.

However, the original musical scale in Europe was obtained by dividing the octave into a series of ratios of the original fundamental frequency. So rather than 164.81 for E_3 in figure 2 (which is a fifth on today's pianos), the fifth was originally simply 3/2 times the fundamental (A_2 in this example) which is 165Hz. The reason the change was made to the tempered scale we use today is to improve the tonal quality when playing in different keys. As western music developed in Europe, keyboards (usually harpsichords and organs) were tuned for a particular group of keys, in order that all the instruments, especially strings, sounded "right" in those keys. So the harpsichord sounded great in those keys,[14] but out of tune in other unrelated keys (say C#). If a guitarist plucks a string on a guitar, and searches for the harmonics by lightly touching the string (without making it touch the frets) he or she will find pure Fibonacci relationships in the frequencies that are connected to the fundamental.

12 D.H. Bailey and J.M. Borwein, "Why Are so Many Mathematicians also Musicians?" *Huffington Post* March 5, 2016. See http://www.huffingtonpost.com/david-h-bailey/why-are-so-many-mathemati_b_9814796.html. Accessed Sept 2016.

13 Music and Mathematics, Wikipedia. See https://en.wikipedia.org/wiki/Music_and_mathematics. Accessed Sept 2016. A modified version of a figure from that article is used in figure 2. Note that this same article discusses interestingly other musical systems which split the octave into 19 notes and the Arabic system which uses 24 notes.

14 G. Meisner, "Music and the Fibonacci Sequence and Phi," May 2012. See http://www.goldennumber.net/music/. Accessed Sept 2016.

Note in Scale	Frequency (Hz) f	Frequency Distance from previous note	Log frequency $\log_2 f$	Log frequency Distance from previous note
A_2	110.00		6.78136	
$A\#_2$	116.54	6.54	6.86468	1/12 = 0.08333
B_2	123.47	6.93	6.94802	0.08333
C_3	130.81	7.34	7.03133	0.08333
$C\#_3$	138.59	7.78	7.11468	0.08333
D_3	146.83	8.24	7.19800	0.08333
$D\#_3$	155.56	8.73	7.28133	0.08333
E_3	164.81	9.25	7.36466	0.08333
F_3	174.61	9.80	7.44799	0.08333
$F\#_3$	185.00	10.39	7.53138	0.08333
G_3	196.00	11.00	7.61471	0.08333
$G\#_3$	207.65	11.65	7.69801	0.08333
A_3	220.00	12.35	7.78136	0.08333

Note in Scale	Equally Tempered Frequency	Calculated Original Frequency	Ratio to Root Octave	Ratio to Octave below/above	Musical Relationship
E_3	164.82	165	3/8 = (3/2)/2/2	3/4 = (3/2)/2	Fifth
F_3	174.62	176	2/5 = (8/5)/2/2	4/5 = (8/5)/2	Augmented Fifth
$F\#_3$	185.00	183.33	5/12 = (5/3)/2/2	5/6 = (5/3)/2	Sixth
A_3	220.00	220	1/2	1/1	Octave below
C_4	261.63	264	3/5	6/5 = (3/5)*2	Minor Third
$C\#_4$	277.18	275	5/8	5/4 = (5/8)*2	Third
D_4	293.66	293.33	2/3	4/3 = (2/3)*2	Fourth
E_4	329.63	330	3/4 = (3/2)/2	3/2	Fifth
F_4	349.23	352	4/5 = (8/5)/2	8/5	Augmented Fifth
$F\#_4$	370.00	367	5/6 = (5/3)/2	5/3	Sixth
A_4	440.00	440	1/1	2/1 1/2	Root Octave
C_5	523.25	528	6/5 = (3/5)*2	3/5	Minor Third
$C\#_5$	554.36	550	5/4 = (5/8)*2	5/8	Third
D_5	587.32	586.67	4/3 = (2/3)*2	2/3	Fourth
E_5	659.26	660	3/2	3/4 = (3/2)/2	Fifth
F_5	698.46	704	8/5	4/5 = (8/5)/2	Augmented Fifth
$F\#_5$	740.00	733.33	5/3	5/6 = (5/3)/2	Sixth
A_5	880.00	880	2/1	1/1	Octave above
C_6	1,046.50	1,056	12/5 = (3/5)*2*2	6/5 = (3/5)*2	Minor Third
$C\#_6$	1,108.72	1,100.00	5/2 = (5/8)*2*2	5/4 = (5/8)*2	Third
D_6	1,174.64	1,173.33	8/3 = (2/3)*2*2	4/3 = (2/3)*2	Fourth

3

2 The western (equally tempered) musical scale — modified from reference 13 **3** The western original musical scale

It is when one looks at the *original* musical scale (figure 3) that Fibonacci numbers appear quite naturally as figure 3 shows. What one can see is that ratios involving 1, 2, 3, 5, 8 are in the main chords and as one moves through octaves, then the number 2 either multiplies or divides these ratios. In figure 3 we are taking the root octave note as A_4 and considering chords in the key of A. As one can see for instance from E_5 the frequency for the evenly tempered convention is very close to 3/2 times the fundamental but not exact, and in fact it is just slightly flat.

Furthermore, the very notes on a keyboard (figure 1) also show connections with the Fibonacci series. There are 13 notes (including the semitones) in any octave and a basic scale is composed of 8 notes, of which the 3rd and 5th notes create the basic foundation of all chords.[15] Moreover, some musical instruments are also made with ratios of lengths corresponding to the golden ratio.[16] Some compositions of the great masters such as Bach, Beethoven, and others have the amount of time in a symphony or concerto spent on one theme compared to another in ratios corresponding to the golden ratio $\Phi = 1.618$. Garland discusses this and many other examples in her book on the Fibonacci series.[17] The beauty of classical western music is based on the Fibonacci series and the golden ratio.

This shows that musical beauty is connected to patterns and structure. Music is a gift from God who delights in order (1 Corinthians 14:40) and heaven has good music (Revelation 5:9).

15 "What Are the Frequencies of Music Notes? Interactive Mathematics." See http://www.intmath.com/trigonometric-graphs/music.php. Accessed Sept 2016.

16 Guy Meisner, "Music and the Fibonacci Sequence and Phi," May 2012. See http://www.goldennumber.net/music/. Meisner shows that good violins are designed to have the ratio of total length to the length of the main sound box to be $\Phi = 1.618$.

17 T.H. Garland, *Fascinating Fibonaccis* (Dale Seymour Publications, 1987), chapter 4, "Fibonacci Numbers in Music and Poetry," p. 33–34.

Elegance in Mathematical Equations

Some remarkable equations which have enormous predictive power

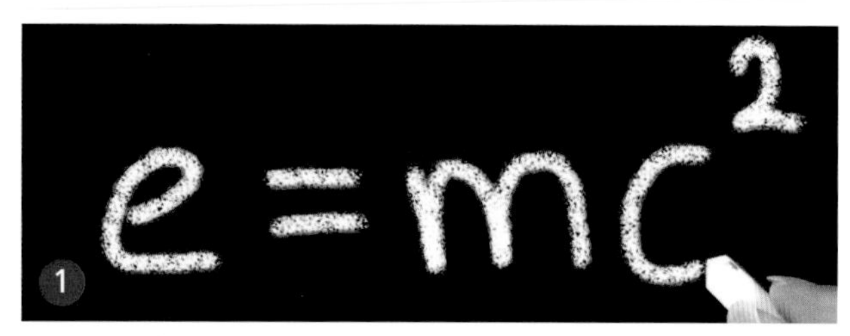

1 Einstein's famous equation connecting mass and energy.

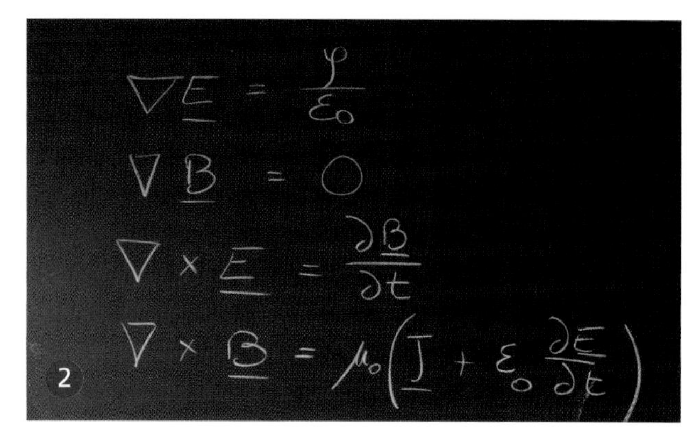

2 Maxwell's equations describing the interaction of electric (E) and magnetic (B) fields

Most of us will baulk at equations! However, it is amazing that some of the most significant physical concepts are built around some of the most elegant equations and the power of mathematics is that it can, rightly used, both guide experimental work and predict profound connections in the sciences.

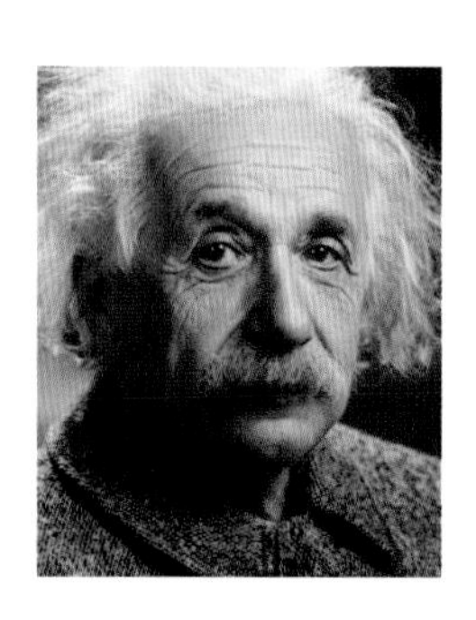

$E=mc^2$

One of the most famous scientists is certainly Albert Einstein who established the connection (figure 1) between energy and mass through the equation $E = mc^2$ where E stands for Energy, m the mass, and c the speed of light.

Deceptively simple, this equation describes one of the greatest discoveries of the 20th century when Einstein (1879–1955) showed that the mass of an object can be described in terms of its energy.[18] What was once of mere academic interest took a more sinister turn by the mid-20th century when the Manhattan project, led by experts at Los Alamos in New Mexico, developed a workable device which converted mass into such energy that had never been known before — nuclear power. Soon the terrifying events of Hiroshima and Nagasaki in August 1945 brought in one of the most powerful weapons ever devised, and effectively ended the war in the Pacific. However, this also led to an arms race and the development of weapons still stored by major political powers today.

This all goes back to the 1905 paper by Einstein. Because the speed of light is so great, this shows that when mass is actually lost under nuclear fission (when the nucleus of an atom loses sub-atomic particles), this causes an enormous amount of energy in the form of radiation (including great heat and other forms of harmful atomic radiation) to be emitted.

Nuclear fission can also be used for peaceful purposes, in particular for creating a clean power supply that does not rely on coal, gas, and other fuels which have been depleting as many nations in the world become large manufacturing centers.

Maxwell's equations

Another profound example is perhaps the most significant set of equations ever. James Clerk Maxwell (1831–1879) was born in Edinburgh and in a short life brilliantly showed the connection of magnetism with electricity. Einstein described

18 A. Einstein, "Does the Inertia of a Body Depend upon Its Energy Content?" *Annalen der Physik*, Sept. 27, 1905.

Maxwell's work as "the most profound and the most fruitful that physics has experienced since the time of Newton."[19]

These four equations (figure 2) describe four important principles concerning the interaction of electric and magnetic fields.

1. The electric flux leaving a given volume is proportional to the charge within that volume (Gauss's Law).
2. The total magnetic flux through a closed surface is zero (Gauss's Law for Magnetism).
3. The voltage induced in a closed circuit is proportional to the rate of change of the magnetic flux it encloses (Faraday's Law).
4. Electric currents and changes in electric fields are proportional to the magnetic fields circulating about the areas where they accumulate (Ampere's Law).

The brilliant step of Maxwell was combining all four laws together and finding a route to solving what is essentially a wave equation (figure 3) at the heart of these major statements of principle. This then led to him being able to predict the speed of such an oscillating wave front.

The speed of the wave was shown to be:

19 P. McFall, "Brainy Young James Wasn't so Daft After All," *The Sunday Post*, April 23, 2006, maxwellyear2006.org. Retrieved March 29, 2013. http://www.maxwellyear2006.org/html/press_coverage.html#Press5. Accessed Sept 2016.

$$c = \sqrt{\frac{1}{\varepsilon_0 \mu_0}}$$

This is the speed of light, which is a function of the properties of space — the so-called electric permittivity ε_0 of free space and the magnetic permeability μ_0 of free space. The wave speed is close to 671,000,000 mph, and it was later measured to be very close to this value. So Maxwell had shown using only mathematical tools that light was a wave, and he had shown what speed this wave should have entirely from mathematical logic! He had also proved, using a succinct mathematical model, that light was an electro-magnetic phenomenon.

James Clerk Maxwell was a committed believer in God and creation. He stated in his inaugural lecture at Aberdeen:

> While we look down with awe into these unsearchable depths and treasure up with care what with our little line and plummet we can reach, we ought to admire the wisdom of Him who has so arranged these mysteries that we can first find that which we can understand at first and the rest in order so that it is possible for us to have an ever increasing stock of known truth concerning things whose nature is absolutely incomprehensible.[20]

Thus, beauty and elegance in mathematics reflects the very wisdom of God.

20 J.C. Maxwell, Inaugural Lecture, Aberdeen, November 3, 1856, *Scientific Letters and Papers*, 1:427.

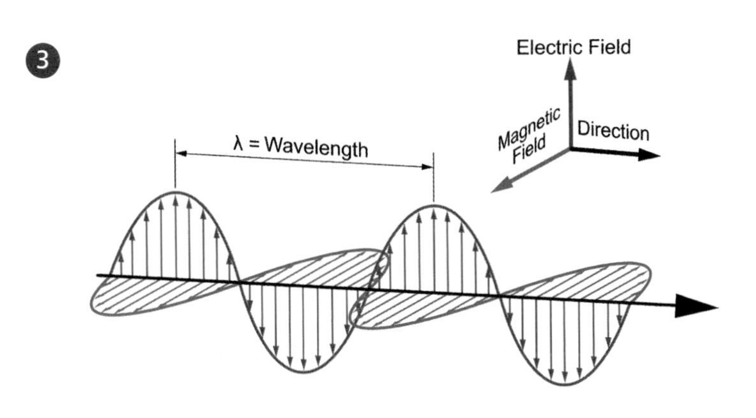

3 Maxwell's wave system of electric fields oscillating in one plane and magnetic fields oscillating in a plane at right angles

10 **Materials** — Inorganic Materials

The world contains a great variety and abundance of non-biological materials that meet all of the needs of mankind in every area of human endeavor.

1 The nickel coin (USA)

2 Titanium fan blades

In this section, non-biological materials are defined as materials that do not come from creatures or plants. They include metals, rocks, minerals, and elements like carbon.

Metals

God has created a wide range of metals that are just right for the varied needs of human activity. There are hard metals like nickel which help to make coins hard (figure 1), soft metals like lead for guttering, lightweight metals like aluminium for drink cans, super-strong metals like titanium for turbine blades (figure 2), and high-conducting metals like copper for use with electricity and heating systems.

One of the most important metals is iron. The durability of iron is illustrated by the Pontcysyllte Aquaduct on the Llangollen Canal in Wales. Built in 1805 of cast iron and 336 yards long, it is still in regular use today (figure 3).

When tiny amounts of carbon are added to iron, it produces a dramatic change in structure and forms steel. Steel is a tremendously useful material because it is extremely strong and tough. It also comes in a great range of types depending on the exact ingredients and processing techniques. For example, when the metal chromium is added to iron and carbon, it forms a rust-free stainless steel which is used for items like cutlery (figure 4). Steel can also be heat-treated to produce special properties like extra hardness or extra toughness.

There are precious (meaning rare) metals such as platinum and palladium that can be used to make catalysts in catalytic converters. These reduce pollution from car exhausts by converting harmful gases into less harmful ones. Other precious metals, like gold and silver, are used for making jewelry and works of art. One special property of gold is that it can be found in the ground in its pure form and does not need to be extracted from ore; in addition, it is the least reactive of all metals, which means that it does not rust or form tarnished surfaces like other metals do. Gold also has the desirable property of

Photo Brian Edwards

3 The cast iron Pontcysyllte Aquaduct is a listed World Heritage Site and carries narrow boats 125 feet above the Dee Valley.

being easily formed or cast into intricate shapes (figure 5). Gold is not just useful for jewelry but is also very useful for medical applications; for example, caps for teeth.

Some metals like uranium and plutonium are extremely dense, with many protons and neutrons packed in their nucleus. Whereas iron has an atomic mass of 56 (30 neutrons and 26 protons), plutonium has an atomic mass of 244 (150 neutrons and 94 protons). This extremely high atomic density results in a very concentrated source of energy which can be released in nuclear reactors to create large amounts of heat. This heat is then used to drive turbines to generate electricity.

4 Stainless steel cutlery

5 Intricate gold bangles

In contrast to nuclear metals, the metal lithium is extremely lightweight. Lithium has an atomic mass of just 7 and is the least dense solid element. Lithium is highly reactive, which means it easily reacts with other substances. This reactivity is harnessed to make powerful lithium batteries. Lithium has several other industrial applications, including lithium grease lubricants and additives for steel and aluminium production.

Remarkably, the abundance of metals closely matches their demand. Aluminium is one of the most useful metals because of its very low density, but it is also the most abundant metal. In fact, it is the third most abundant element on Earth after oxygen and silicon. Iron is the fourth most abundant element, which is ideal because, like aluminium, steel (which is made from iron) is an extremely useful metal.

6 The Taj Mahal is clad in white marble. *(Photo from Wikimedia)*

Another remarkable feature of some metals such as iron, zinc, copper, manganese, and chromium is that they are important for living creatures and plants. For example, humans and animals need iron for oxygen transport in the blood. In a similar way, iron helps to carry important elements through a plant's circulatory system, which is essential for the production of chlorophyll. It is amazing that God can use a single element like iron for such diverse functions as enabling the formation of steel and transporting elements within living creatures!

Rocks

Rocks may look rugged and simple from a distance, but they actually contain much sophisticated design that is useful to mankind. Rocks meet so many of our varied needs in the construction industry and even produce some gems of beauty.

Igneous rocks are formed when magma rises from the depths of the Earth to the surface and then cools. Igneous rocks are composed of one or more minerals, and minerals are composed of one or more elements. For example, granite is a type of rock that is made up of the minerals quartz, feldspar, and biotite. The mineral quartz is made of the elements silicon and oxygen. Granite is an extremely hard and beautiful building material due to its distinctive patterns and colors. Granites are mostly white, pink, or grey, depending on the mineral content.

7 Pottery being molded before being heated

Another type of rock is *metamorphic rock,* which is the result of a transformation of an existing rock. Marble is an example of a beautiful metamorphic rock that is used in the building industry. Some famous structures have been built with white marble, including the Taj Mahal (figure 6) and the ancient temple of the goddess of Artemis at Ephesus (destroyed in A.D. 262). Slate is another metamorphic rock which is commonly used as a tiling material because it can be cut easily into flat sections.

A third major category of rock is that of *sedimentary rocks*, such as sandstone and limestone, which are formed from small particles like sand.

Minerals

There are more than 3,500 different types of minerals that have been identified and these have a great range of properties. The structure of a mineral depends on the temperatures and pressures experienced during its formation.

The rarest and most beautiful minerals are called "gems." The most highly prized gems are diamonds, rubies, sapphires, and emeralds, but there are many others. It is interesting to note that gemstones come

8 Bradley Wiggins on a CFRP bike (Tour de France 2012)

9 A cut diamond

in virtually every color, showing that God loves variety and beauty.

Sand is composed of one or more minerals and is formed when rock is broken down by weathering and water action. Sand is used in many types of building materials such as concrete, mortar, plaster, asphalt, and bricks. Perhaps the most enjoyable function of sand is to form beautiful sandy beaches!

Clay particles are also composed of one or more minerals but are much finer than sand particles. In fact, they can be 1,000 times finer. Clay is used to make pottery (figure 7) by heating it in a furnace. When heated with powdered limestone, clay forms cement, which is an extremely important binding material.

Silicon

Silicon is the second most abundant element in the Earth's crust after oxygen, and is used widely in industry. Silicon occurs mostly in the form of a mineral called silica (silicon dioxide) which is the most abundant mineral in the Earth. It is commonly found in the natural form of quartz, and throughout history this has been used to make jewelry. Silica is now used extensively to make glass, optical fibers, and microelectronic components.

Carbon

Carbon is an important element used in human technology. One of the most common ways of harvesting carbon is directly from coal. Carbon has strong bonds called "covalent bonds" that make it extremely strong in certain directions. Carbon atoms can be arranged in different ways leading to remarkably different properties. It can make both extremely hard materials and extremely soft materials. Diamond and graphite are two examples of carbon in its pure form. Diamond is the hardest material on Earth and is made up of one giant molecule of carbon atoms.

By contrast, graphite is a very soft material used for things like pencils and lubricants. Engineers often use carbon in a composite material called CFRP (carbon fiber reinforced plastic) which is used to make structures that are both strong and lightweight such as racing bike frames (figure 8).

Graphite is a good conductor of electricity and is used by engineers for electrical contacts. On the other hand, diamond does not conduct electricity.

Another striking contrast between different carbon materials is their appearance: whereas graphite and carbon fiber are black and dirty, diamond is transparent and incredibly pure (figure 9).

Carbon is also an important element in living organisms. Complex organic molecules like proteins are made up of carbon bonded with other elements, especially oxygen, hydrogen, and nitrogen. For a simple element to have such a variety of forms and uses clearly reveals the wisdom of the Creator.

Organic Materials

The world contains an abundance of biological (organic) materials that are used extensively in human industry and recreation.

1 The *Mary Rose*, the flagship of Henry VIII which sank off Portsmouth in 1545, was built from British oak.
2 A thatched cottage at Corfe Castle Village in Dorset

In this section, biological materials are defined as materials that come from living or dead organisms such as wood, cotton, wool, silk, and fossil fuels.

Wood and grass

The range of properties of different woods is truly remarkable. Woods vary greatly in properties such as density, hardness, toughness, water-resistance, and strength. The wide range of properties means that woods are useful for a variety of applications. For example, a high-density wood like lignum has a density about ten times higher than that of balsa wood. Lignum is so hard that it can be used for bearings in ships.

Fast-growing trees like pines and cedars are convenient for use in the building industry, whereas woods like cypress are resistant to rot and are therefore excellent for outdoor structures. Some trees produce rubber which has many useful applications.

Willow and ash are strong and pliable for applications requiring toughness, like tool handles. In fact, willow and ash are so tough that they are ideal for cricket and baseball bats. Some woods are particularly beautiful when planed flat, such as teak and oak, which makes them attractive for furniture. Oak was used for many centuries as the main material for the sailing ships that were used to build empires and pioneer trading routes. It is estimated that around 600 large oak trees were felled for the construction of the *Mary Rose* (figure 1).

However, it is not only trees that are useful for structures. Bamboo is a type of tropical grass and yet it is very stiff and can be used as a lightweight structure; it is still used for scaffolding in some countries today. Even straw and water reeds can be used as roofing materials to make "thatched" roofs for buildings (figure 2). In past centuries, thatched roofs were very common, providing excellent insulation and effective waterproofing.

Cotton and wool

Cotton is an important natural fiber that has been used extensively to produce many types of clothing and cloth. It is a soft fiber that grows on cotton plants in clumps called "bolls" which cover and protect the seeds (figure 3). These bolls can be harvested by hand or machine and turned into commercial cotton. There are many plants that cover their seeds with simple casings, so it is remarkable that the cotton plant produces a covering of cotton that happens to be immensely useful to us.

Some animals produce wool and fibers that are also ideal for clothing. Sheep produce large amounts of wool that has high insulation due to the way it traps air between its fibers. Cashmere goats (figure 4) produce particularly smooth fibers which are used for fine clothing. Engineers have made artificial fibers like polyester for high quality clothing, but their quality never matches that of the natural fibers of wool which, unlike synthetics, is also perfectly biodegradable. In addition, whereas natural fibers like cotton absorb moisture, synthetic fibers attract our oily sweat instead, which is why synthetic clothes are smellier after our exercise!

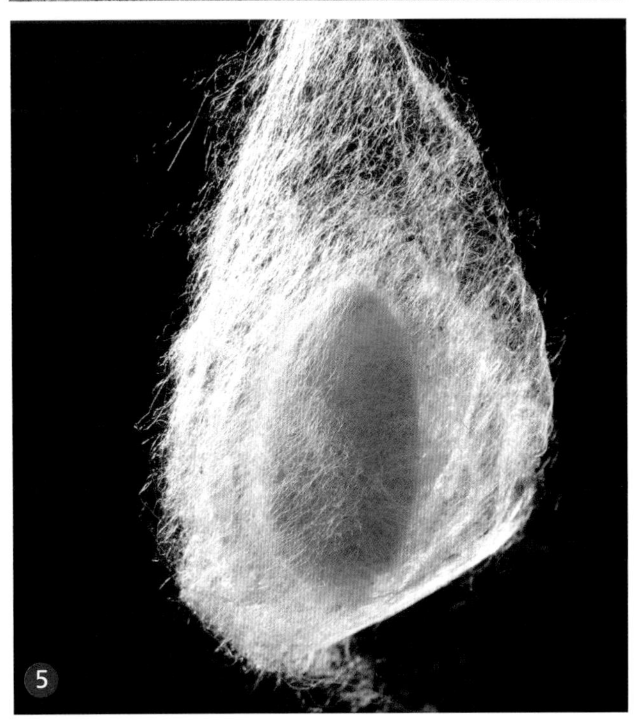

3 Cotton bolls

4 Cashmere goats in Zanskar, North India

5 A silkworm net. The Chinese have cultivated silk from silkworms for thousands of years.

Silk

Some creatures produce silk, which is an extremely strong and smooth fiber made from protein. Silk is often harvested from the silkworm, which is a caterpillar usually of the *Bombyx mori* moth. The silkworm produces silk to form a cocoon which covers the pupa (figure 5). To construct its cocoon, the silkworm produces over 1,000 yards of filament! Billions of silkworms are cultivated in order to produce silk for the silk industry. They are fed on leaves from the white mulberry tree, as this produces high quality silk. It is remarkable how simple leaves can be turned into silk by a caterpillar! The silkworm is an insignificant little creature and yet it can produce the most incredible material that is useful for clothing.

Pearls

A pearl is beautiful because of the multiple layers of a smooth material called "mother of pearl" (or nacre) that causes optical coloring. The optical coloring occurs when light reflects off different surfaces of nacre, causing colors that have a deep luster and iridescence. Natural pearls are so beautiful that people will pay high prices for them. Yet each pearl starts off as a piece of grit.

A pearl is produced when this tiny piece of grit enters a mollusk like an oyster. The creature responds by covering the grit with layers of nacre. So many layers are added that the pearl gradually grows in size and becomes round and beautifully colored
(figure 6).

Feathers

Feathers are the lightweight structures used by birds for flight (see page 52–55 for the remarkable design of feathers for flight). They have always been of great use to mankind as well. The quills of large birds like geese were one of the most common writing implements before the invention of ink pens. In fact, the word "pen" is derived from the Latin *penna*, meaning feather. Quills are well suited to writing with ink (figure 7) because of the way the feather does not bend when held and the hollow interior is ideal for storing ink. Bird feathers have also long been used for the fletches of arrows to keep them straight in flight. Colorful feathers such as those of pheasants were commonly used for fly fishing and are still used for this today.

Feathers also make good insulation material, especially down feathers which are the fluffy undercoat that keeps birds warm. Down feathers of geese and ducks are particularly used for pillows and duvets. They are well suited for this because they do not have hard shafts; they have many other desirable properties: long-lasting, soft, breathable, heat-retaining, and easy to compress.

7 A writing quill

6 A pearl and oyster shell

8 Oil pump rigs

Oil, coal, and gas

The world has huge reserves of oil, coal, and gas on land (figure 8) and under the oceans. The majority of crude oil is used for fuel for transport and heating. However, significant amounts are used to make plastics, chemicals, and lubricants. Depending on the additives and type of processing, there are many forms of plastic that can be produced with a range of properties that suit various applications.

Coal is also a convenient material for fuel and a source of carbon. Gas is used both for fuel (figure 9) and for converting into plastics. Most of the oil, coal, and gas deposits in the Earth were probably formed during the global Flood at the time of Noah. It is known from modern processing that carbon fuels do not need vast amounts of time for their formation.

9 Currently, natural gas provides 80% of heating in the UK and 30% of electricity generation.

Designed for mankind

This abundant and varied provision of inorganic and organic materials, each precisely what is needed for the development of man-made technology, is compelling evidence that planet Earth was designed purposefully for the human race. Despite the fact that fossil fuels are a result of judgments like the worldwide Flood, God in His wisdom has given them a remarkable usefulness for mankind.

The Bible teaches that God formed the Earth to be inhabited: "This is what the LORD says — he who created the heavens, he is God; he who fashioned and made the earth, he founded it; he did not create it to be empty, but formed it to be inhabited" (Isaiah 45:18; NIV).

⓫ Mankind — The Brain

Of all the wonders of the human body, the most wonderful must surely be the brain, because of its unique intellectual capabilities of understanding, imagination, and appreciation of beauty.

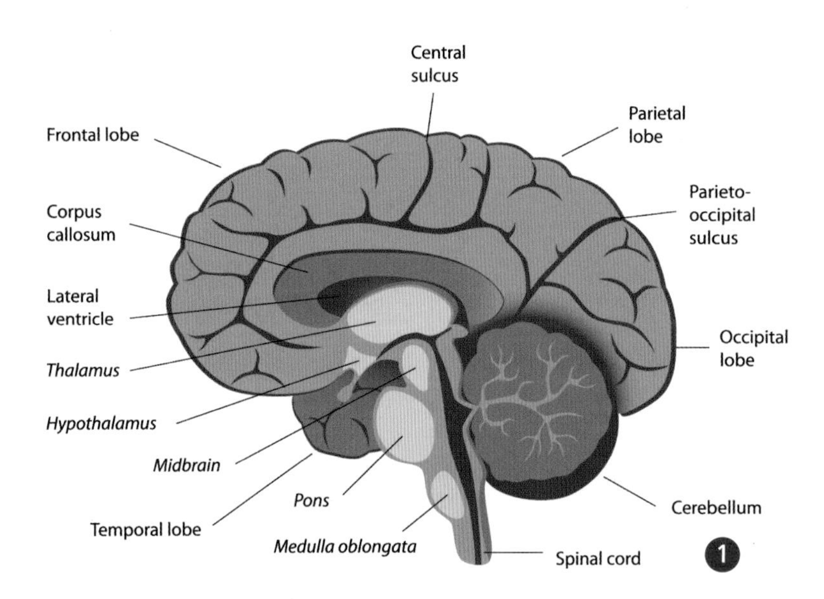

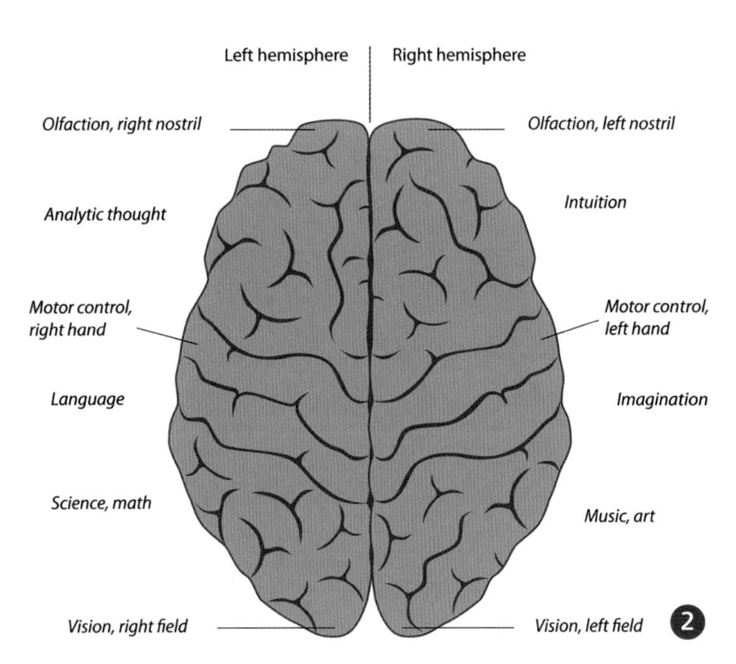

1 The main parts of the brain
2 Some of the functions of the cerebral cortex

The brain is a relatively small part of the body, weighing around 3 lbs. for a typical adult. However, even though the brain represents only 1.8% of the bodyweight, it uses approximately 20% of the body's oxygen, 25% of the body's glucose, and 15% of the blood supply. The reason for this high energy demand is that it is a powerful computer with an immense capacity for processing information. Humans have a very large "brain weight" to "body weight" ratio of around 1.8%. This is far higher than any other animal in creation and at least five times larger than for apes, showing that humans are special beings.

A unique brain

Figure 1 shows the main parts of the human brain. The cerebral cortex is particularly large in humans and the source of unique human abilities such as thinking and planning. In general, the right hemisphere controls the left side of the body and the left hemisphere controls the right side of the body. Figure 2 shows how different areas in the cerebral cortex have specific functions like analytical thought, music, art, imagination, intuition, sensing, and muscle control.

The **medulla**, along with the spinal cord, controls a wide variety of involuntary motor functions such

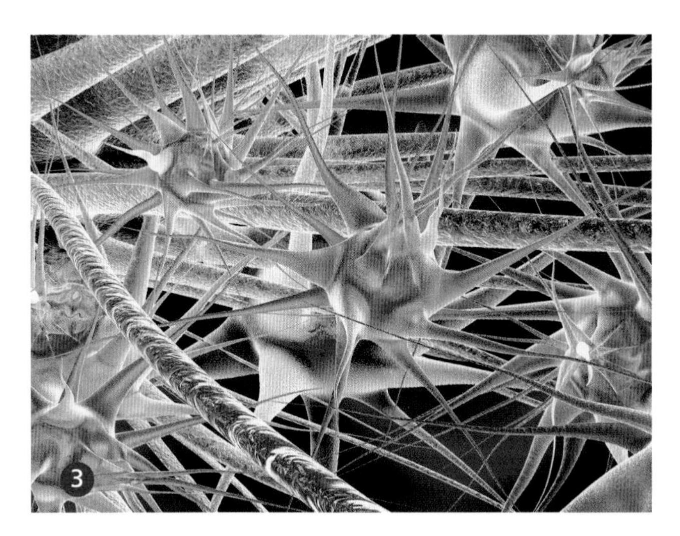

3 Artist's impression of connections in the brain

4 Humans have the unique ability for analytical thought.

as heart rate and digestive processes. The **pons** contains neurons that control voluntary acts such as sleep, respiration, swallowing, bladder function, eye movement, and facial expressions. The **hypothalamus** is engaged in involuntary or partially voluntary acts such as eating and drinking and the release of hormones.

The most complex computer in the world

A healthy human brain has something like 100 billion nerve cells (called neurons) which communicate with one another via synapses (figure 3). Each individual nerve cell can have thousands of connections, which means that there are hundreds of trillions of connections in the brain. The synapses function like computer microprocessors with memory storage and information-processing capabilities.

The neuron cells generate electrical signals that travel to other neurons. When a pulse of electricity reaches a synapse, it causes a neurotransmitter chemical to be released so that the electrical signal is passed on to the next neuron. Activities such as thinking and learning involve millions of electrical signals traveling simultaneously between interconnected networks of neuron cells. New brain connections are created every time a person forms a memory or learns something new.

The staggering complexity of the human brain is far greater than any computer made by man. Studies by brain researchers at Stanford University found that a single human brain has more switches than all the computers in the world combined.[1]

Unique abilities

The human brain has an amazing ability to analyze (figure 4) and have rational thought (see pages 142–143 for more details). Humans also have a unique ability to make rational decisions and have emotional feelings. In contrast, animals follow their instincts or, if they are trained, follow their master's instructions. This shows that humans are made in God's image and have been designed to be stewards of creation.

The human brain also has a unique ability to appreciate beauty. In contrast, animals never stop to look at a beautiful sunset or to listen to birdsong. It should be noted that a person's identity is defined not just by their physical brain but also by their invisible soul. The connection between the brain and the soul is not something that can be explained by us but it nevertheless is an important reality.

1 http://www.cnet.com/uk/news/human-brain-has-more-switches-than-all-computers-on-earth/.

The Nervous System

The nervous system is a vast communication network that connects every part of the body to the brain, enabling the body to be moved with precision and to sense the surroundings.

We are often unaware of the many functions that our nervous system is carrying out each moment of our lives. During daily living, billions of nerve impulses are traveling throughout the human brain and nervous system. This system controls both voluntary functions, like muscle movements, and involuntary functions, like blood circulation and digestion. Most of the control of the body is coordinated by the brain, although some reflex reactions happen in the spinal cord.

The central nervous system

The central nervous system consists of the brain and spinal cord (figure 1). The spinal cord contains millions of individual nerve fibers like an electrical cable with millions of minute separate wires. In adults, the spinal cord is typically 15 to 19 inches long (40 to 50 cm) and 0.4 to 0.6 inches (1 to 1.5 cm) in diameter. The spinal cord passes down through the backbone, which is called the vertebral column. The vertebral column typically contains 33 vertebral bones that provide protection for the spinal cord.

The spinal cord branches off into 31 pairs of nerve roots which exit the spine through small openings on each side of each vertebra. The 31 nerves in each area of the spinal cord connect to specific parts of the body. There are four main sections in the spinal cord: cervical, thoracic, lumbar, and sacral. The nerves of the cervical section go to the upper chest and arms. The nerves of the thoracic section go to the chest and abdomen.

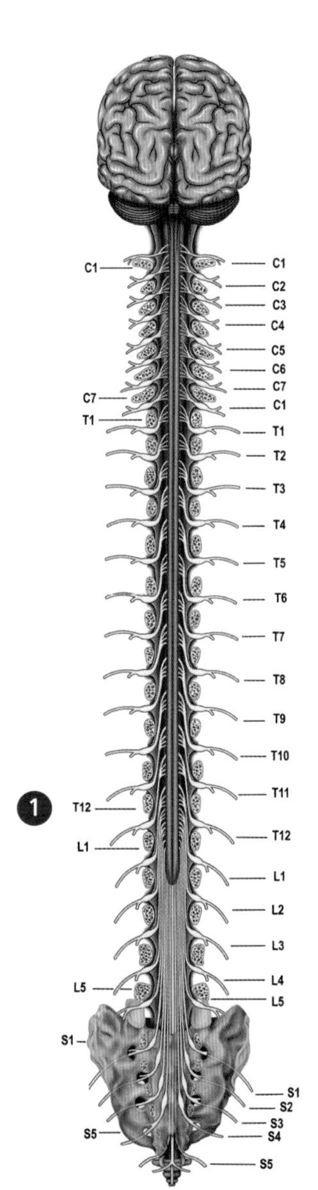

1 The spinal cord branches off into 31 pairs of nerve roots.

2 The peripheral nervous system consists of nerve cells outside the brain and spinal cord.

The nerves of the lumbar and sacral sections go to the legs, pelvis, bowel, and bladder.

The peripheral nervous system

The peripheral nervous system branches out with incredible complexity and precision to reach every part of the body as shown in figure 2. The total length of the peripheral nerves in the human body comes to around 93,000 miles.

Motor nerves carry signals to organs and muscles, whereas sensory nerves carry signals to the brain. Some nerves are called mixed nerves because they send signals both ways. Sensory nerves transmit information about senses such as touch, sight, sound, smell, taste, temperature, and pain.

Dendrites collect signals
Axon passes signals
Neuromuscular junction
Myelin
Muscle fiber

3 The structure of a motor neuron cell

Electrochemical waves

Nerves contain neurons that are specially designed to transmit signals in the form of electrochemical waves. A typical neuron consists of a cell body (soma), dendrites, and an axon as shown in figure 3. The axon has the function of carrying a signal over a distance and can vary greatly in length. Sensory neurons can have axons that run from the toes to the posterior column of the spinal cord which is a total distance of over 4.9 feet in adults. No other type of cell is that long.

An integrated body

One of the amazing wonders of the human body is how all the different sub-systems of the body are integrated with precision and elegance (figure 4). Despite the immense complexity of bringing together such diverse sub-systems, the body is able to function efficiently and reliably. Such a complex integrated mega-system requires the foresight and planning of a Being of infinite understanding.

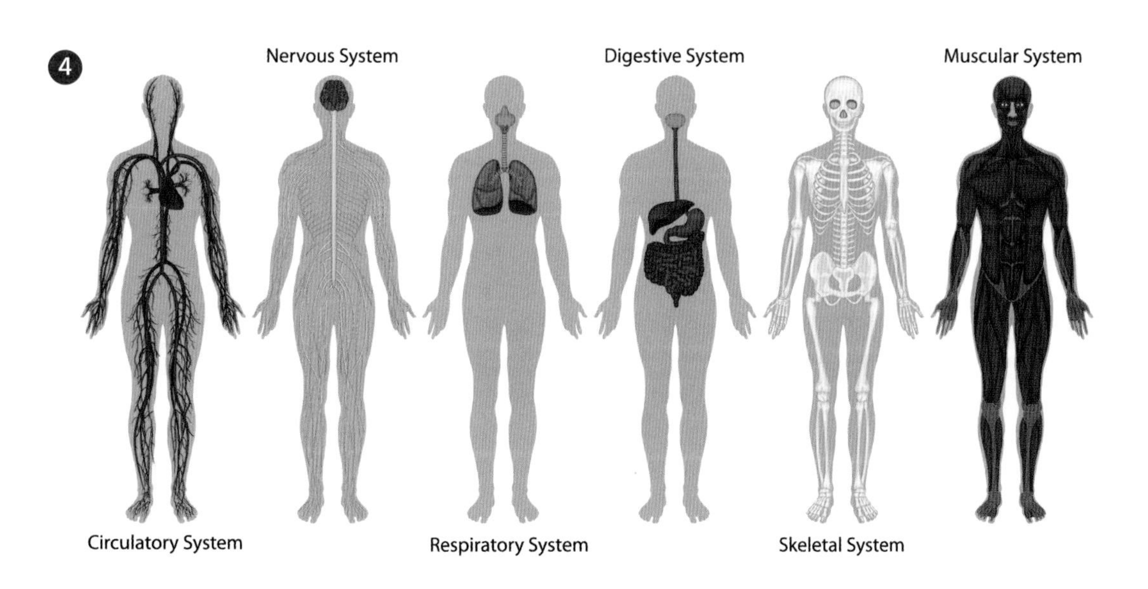

4 Integration of different body systems

The Heart and Lungs

The human cardiovascular system contains an incredibly intricate arrangement of blood vessels that continuously carry out the vital transport of blood around the whole body.

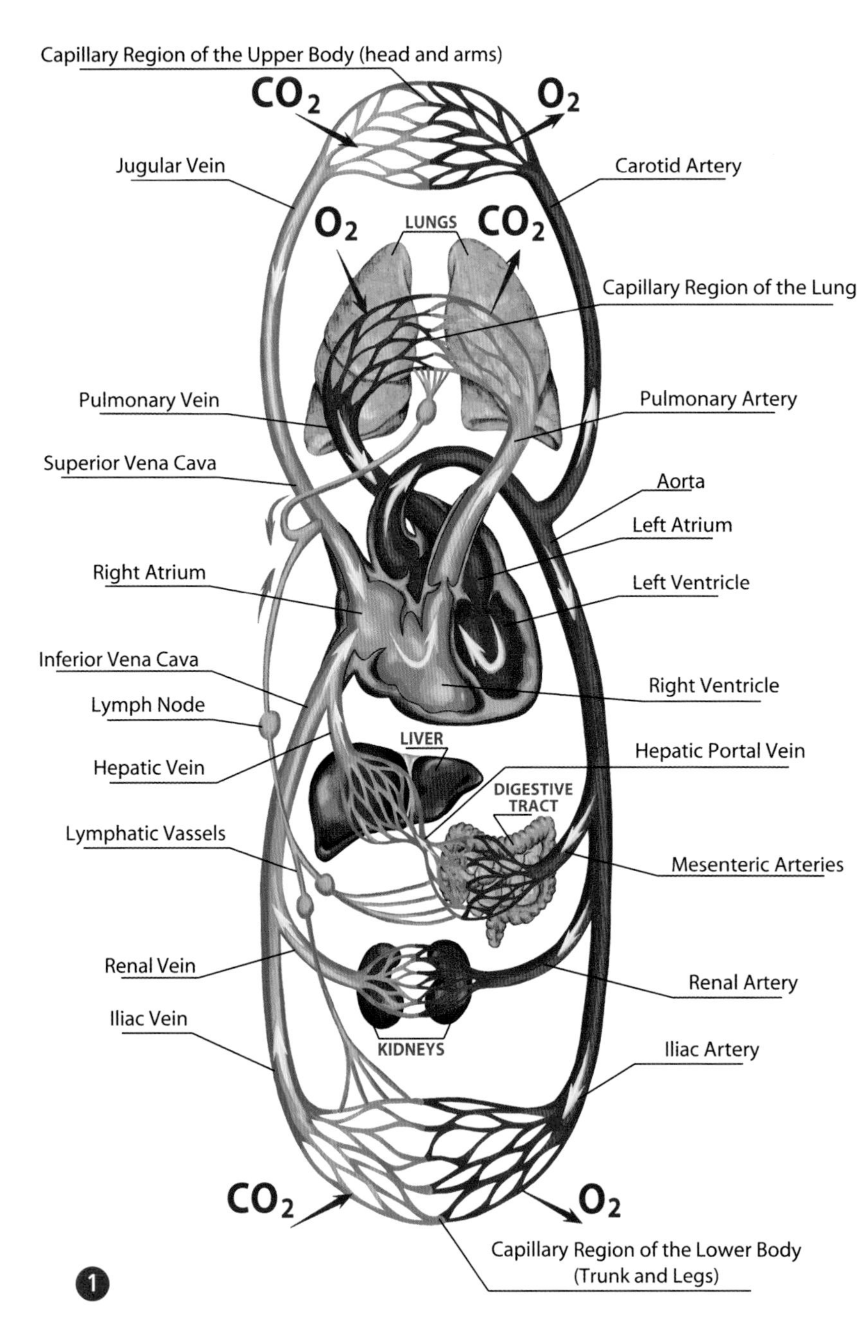

1 Humans have a double circulatory blood system.

The human cardiovascular system contains heart, blood, lungs, and a vast network of blood vessels. The cardiovascular system has several vital functions, which include transporting oxygen and nutrients around the body, removing waste, cooling, heating, healing, and fighting infections.

A double circulatory blood system

The heart pumps blood around two separate circuits as shown in figure 1. In the "lung" (or pulmonary) circuit, in order to get oxygen, the heart pumps blood to the lungs and back. In the "body" (or systemic) circuit, the heart pumps blood around the body.

The heart is able to pump blood around two circuits because the heart is itself a double pump. Having two pumps in the heart is a perfect design because it enables the lung circuit to operate at low pressure and the body circuit at high pressure. This is ideal because gas exchange requires low pressure whereas pumping blood through the long body circuit requires high pressure. The two pumps are sometimes referred to as the right heart (lung circuit) and left heart (body circuit).

The heart — super-powerful

The human heart (figure 2) is a muscular organ about the size of a large orange and typically weighs around 10.5 ounces. The heart has four chambers as shown in figure 2. The two upper chambers (called atria) are where blood returns to the heart from the two circuits. The lower chambers (called ventricles) are where blood is pumped around the two circuits.

Despite its relatively small size, the human heart has a phenomenal power output and endurance. During rest and sleep the heart beats at around 60–100 times a minute, rising to 220 beats per minute during exercise in order to increase the supply of oxygen to the muscles. The human heart beats approximately 100,000 times a day, 35 million times a year, and 2.5 billion times over an 80-year lifespan!

The double circulatory system is a clear example of irreducible complexity, because the system can only function properly if all the parts of the two circuits are in place, including the double pump in the heart. If any of the major arteries like the aorta become completely blocked, then it is always fatal unless there is medical intervention. This emphasizes that complex plumbing cannot evolve step-by-step.

The lungs — super-efficient

The purpose of the lungs is to bring oxygen into the body and to remove carbon dioxide. The lung operates like the inflating of a balloon (figure 3). After entering the nose or mouth, air travels down the trachea or "windpipe" and then into a vast network of airways. The trachea divides into left and right breathing tubes called bronchi. These breathing tubes continue to divide into smaller and smaller tubes called bronchioles. The bronchioles end in tiny air sacs called alveoli. There are over 300 million alveoli in healthy lungs.

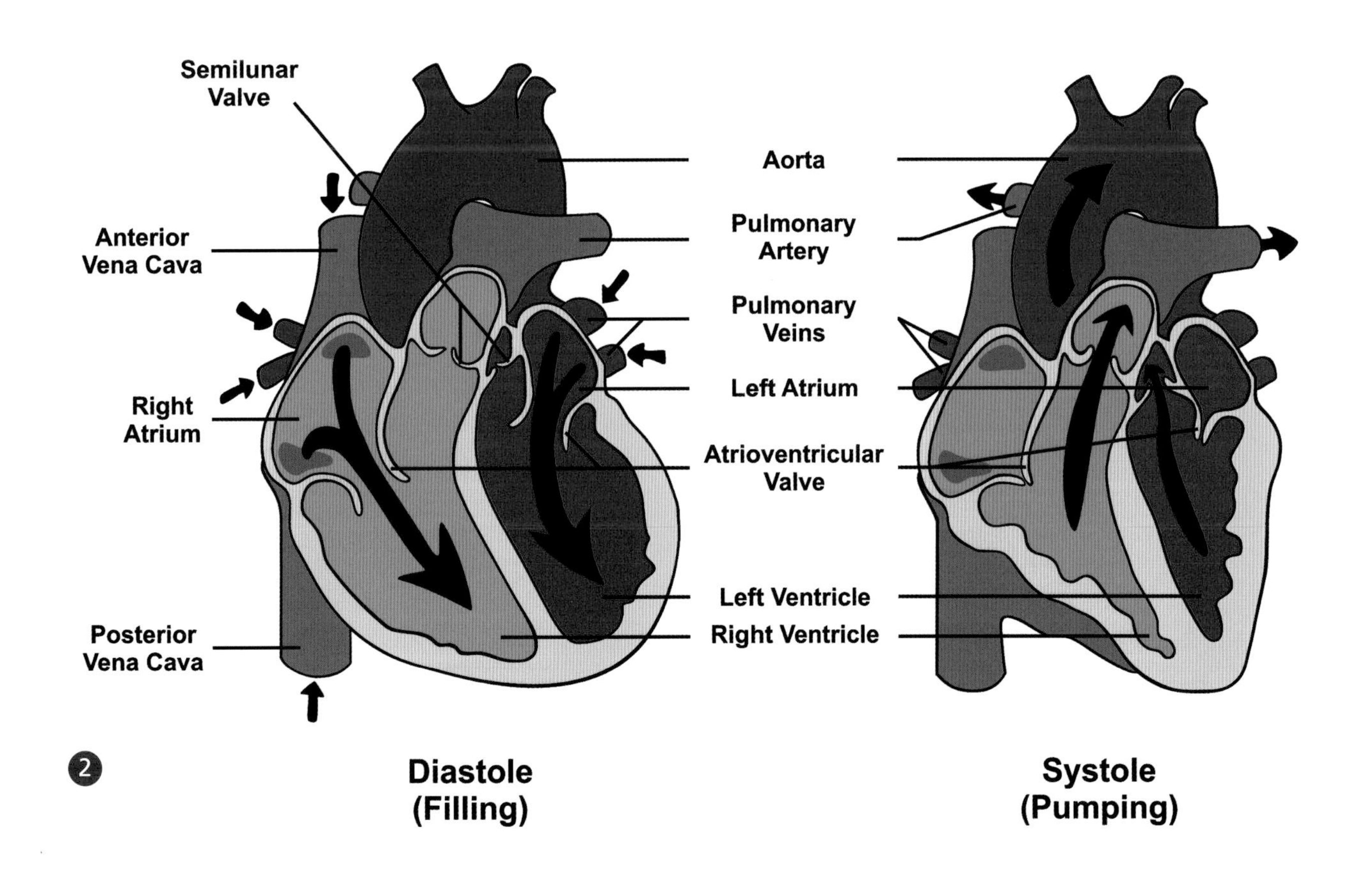

2 The human heart is a double pump

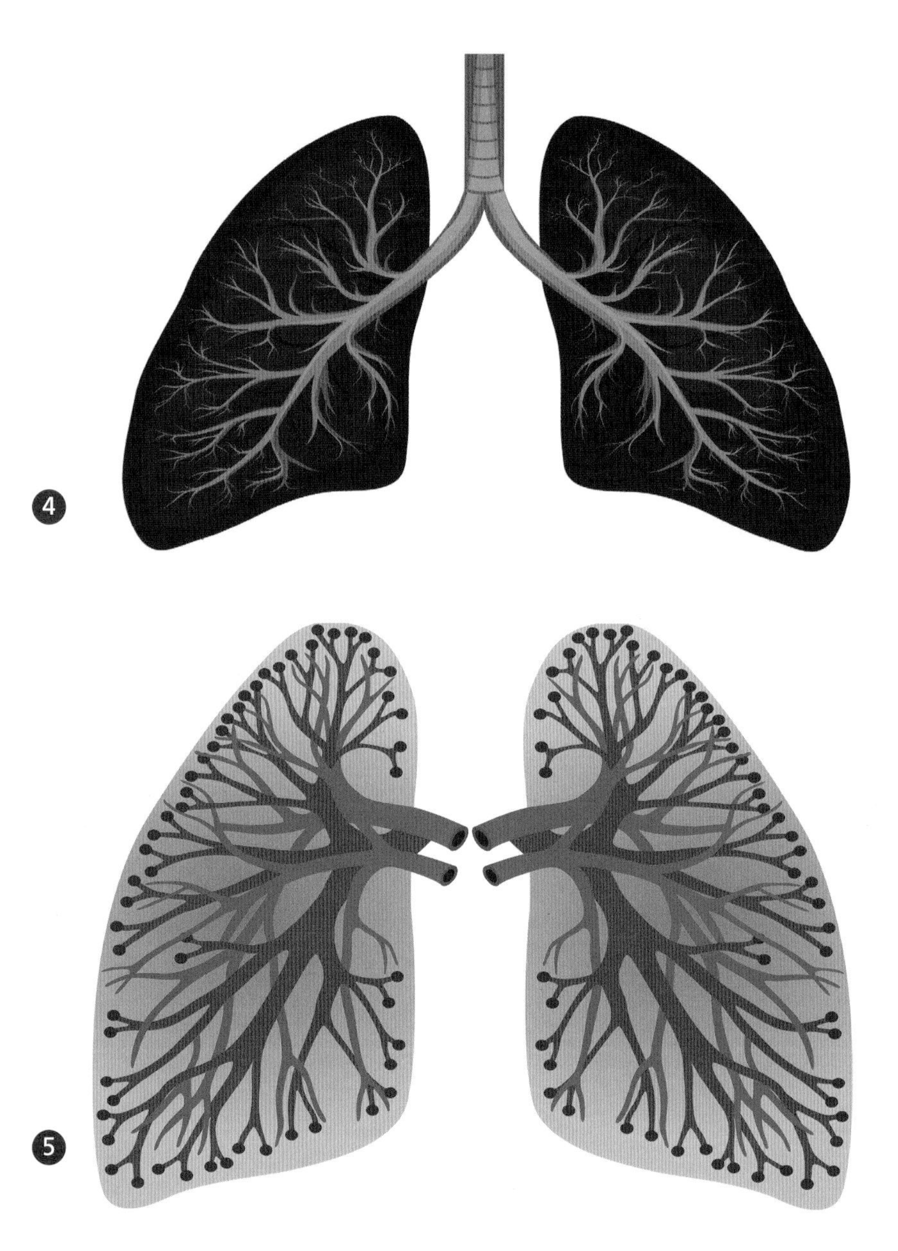

3 The lung operates like the inflating of a balloon. **4** The airway network in human lungs
5 The blood vessel network in human lungs

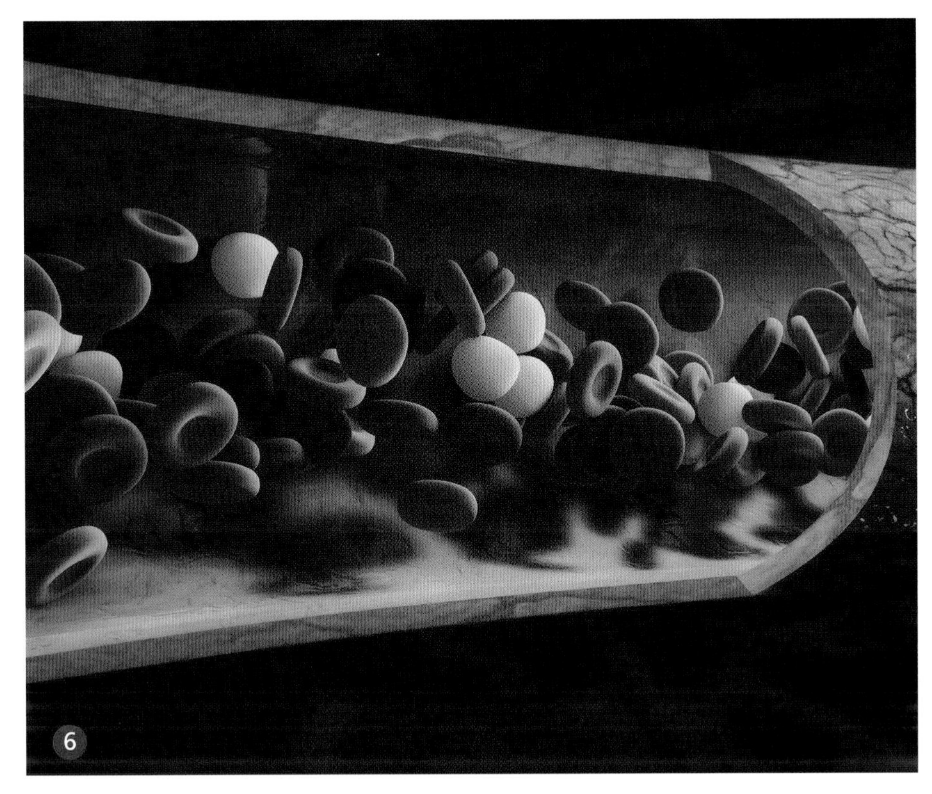

6 Blood cells within a blood vessel **7** The human body requires very fast gas transfer in and out of the blood for all movement.

The air sacs are surrounded by blood capillaries so that oxygen can diffuse into the blood and carbon dioxide can diffuse into the air. Amazingly, the branched network of airways is perfectly integrated with the network of blood vessels (see figures 4 and 5) so that gas exchange happens efficiently between air and the blood.

If all the alveoli were laid out flat, they would cover an area the size of a doubles tennis court. The enormous surface area of the lungs makes them super-efficient.

The blood — life-giving liquid

The average adult has around 8.7 pints of blood in their body. Blood circulates around the body in approximately one minute, which means that in one 24-hour period the blood circulates around the body about 1,500 times.

Blood is an amazing substance that performs multiple functions including:

1. conveying nutrients to cells
2. carrying waste from cells
3. cooling and heating
4. dealing with infection
5. forming blood clots to prevent bleeding after an injury.

No wonder the Bible comments that "The life of every creature is its blood" (Leviticus 17:14)!

All cells in the body need oxygen and glucose for energy production, and these are provided by the blood. Other items carried by the blood include hormones, waste products (e.g., urea, carbon dioxide) and products of digestion (e.g., sugars, fatty acids, and amino acids). Blood also provides cells with water.

There are three types of blood cell: red blood cells, white blood cells, and platelets (figure 6). Red blood cells carry oxygen to body cells and carbon dioxide away from body cells. They are tiny and flexible, allowing them to flow through very small capillaries.

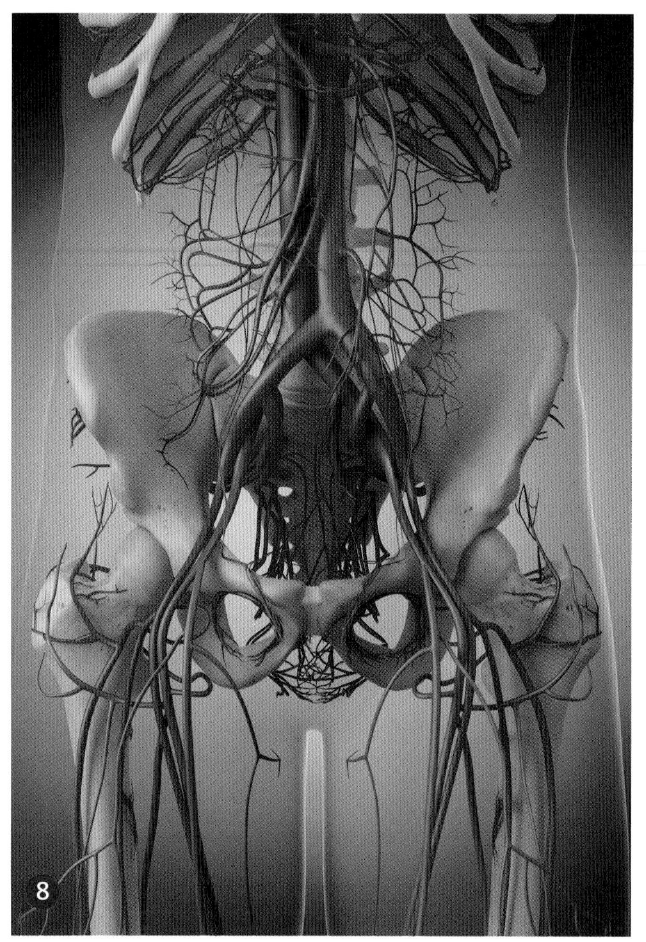

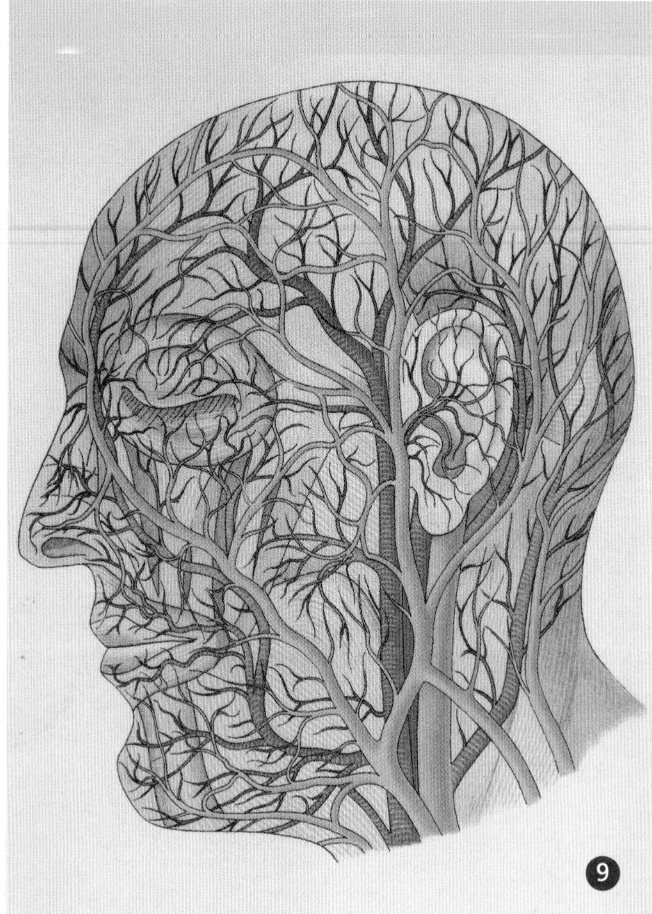

The main function of white blood cells is to protect against disease. They engulf bacteria and produce antibodies (proteins that identify and neutralize harmful bacteria and viruses). So all activity (figure 7) is vitally dependent on blood flow.

Platelets are fragments of cells which initiate the blood-clotting mechanism. If there is a break in the blood vessel wall, the platelets stick to it and change shape to plug the gap.

The blood vessels — an intricate network

Figures 8–10 provide a glimpse of the staggeringly complex blood vessel network in different parts of the body. Blood leaves the heart through arteries which have highly elasticated, thick muscular walls to withstand the high pressure and help pump blood. Blood progresses from arteries into many smaller vessels known as arterioles and from these into the smallest vessels called capillaries.

Capillaries are tiny blood vessels which have a wall thickness of only one cell, enabling them to pass substances quickly and easily between the blood and body tissues. After supplying cells with oxygen and other substances, and receiving their waste products, blood returns to the heart through tiny vessels called venules and then the larger veins. At this point it is at low pressure and is moving against the force of gravity, so the veins contain valves to stop it from flowing back down.

As well as being incredibly intricate, the blood vessel network also has special devices to help alter the flow of blood to different parts of the body.

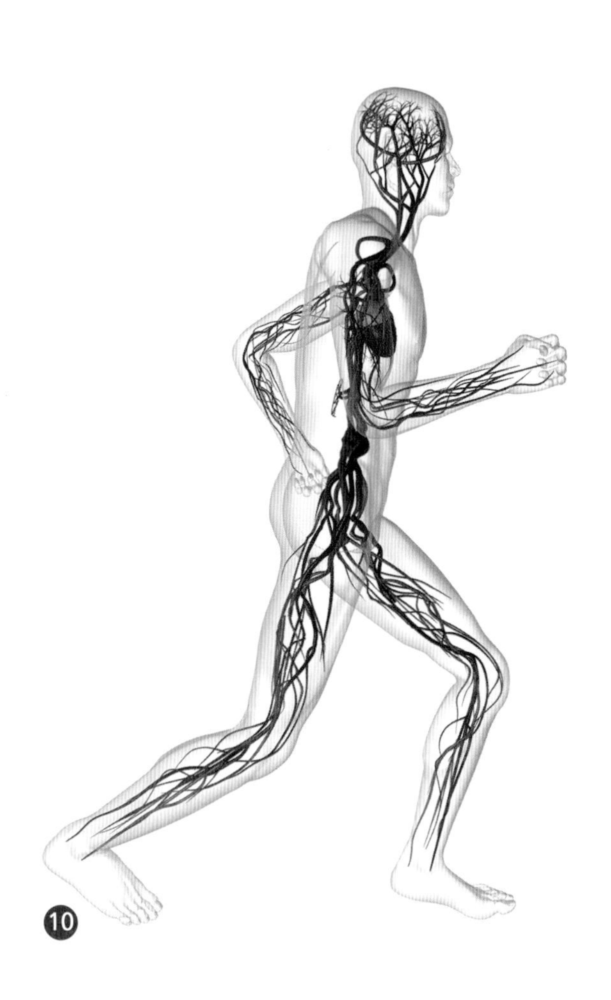

(Note: for figures 8–10, arteries are shown in red and veins are shown in blue.)

8 Medically accurate illustration of the circulatory system — abdomen

9 Blood vessels in the head

10 Blood vessels in the body

11 During exercise as in running sports, blood flow to muscles is increased.

Blood flow to capillaries is controlled by valves called sphincters, positioned where they originate from the arteriole. When the sphincters shut, the blood flows through shunt vessels instead. (A shunt vessel is a blood vessel that links an artery directly to a vein, thus allowing the blood to bypass the capillaries when this is required.) This helps the body conserve heat in cold environments by restricting the flow of blood to extreme parts of the body. This is why the hands and feet feel cold when they are exposed to cold temperatures. During exercise (figure 11), blood flow to muscles is increased and blood flow to certain organs (gut, liver, kidney, spleen) is decreased.

The total length of the blood vessel network in an adult human is around 60,000 miles. When you consider that the body contains about 100 trillion cells and that the vascular system must get blood to within around 0.004 inches of every single cell, you realize that the system has unimaginable complexity and precision!

The amazing sophistication of the blood vessel network in the human body is a clear illustration of the Psalmist's conclusion: "I will praise you for I am fearfully and wonderfully made" (Psalm 139:14).

The Muscles and Skeleton

The human musculoskeletal system gives humans unique skills that enable them to be creative beings and stewards of creation.

1 The human musculoskeletal system

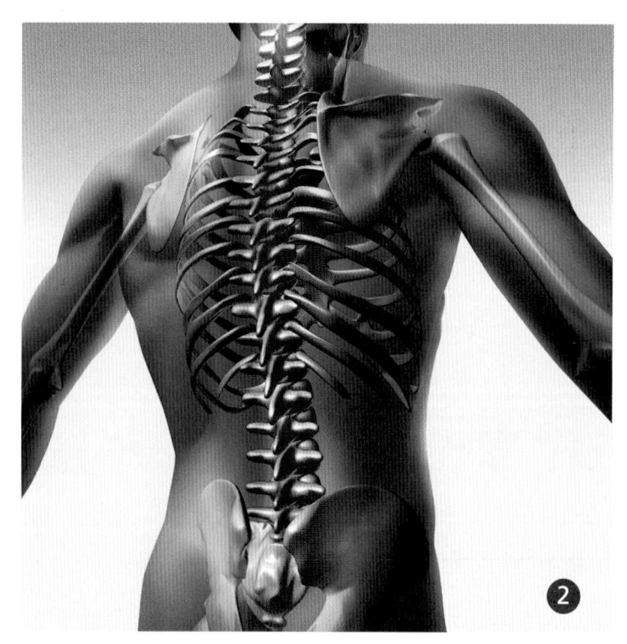

2 The human back

A skillful body

The human musculoskeletal system (figure 1) enables the body to move with skill and precision. The human body has around 210 bones that give it rigidity. These bones are held together with hundreds of ligaments to form free-moving joints. There are around 650 muscles that connect to bones via tendons and enable a myriad of movements from head to toe. The whole system is integrated with blood vessels and nerves in an incredibly complex assembly. But despite the immense complexity, the human body is elegantly and beautifully shaped.

A strong back

The human spine consists of 24 very strong bones called vertebrae. These are kept securely in place because they stack on top of each other in a similar way that drinking glasses stack (figure 2). There is also a soft cartilage disc between each vertebra to reduce wear and tear and provide cushioning.

Humans have a back with a unique S-shape when viewed from the side. This is an ideal design because it gives an upright stature at the same time as avoiding a structure that is too stiff along the spine. The spine is able to deform under load because the S-shape can squash like a spring.

A precision knee joint

The human knee joint is a rolling joint because of the way the femur (thigh) bone rolls over the tibia (lower leg) bone. However, the joint also has a linkage mechanism in that there are two cruciate (cross-shaped) ligaments that form a 4-bar mechanism as shown in figure 3. This is an extremely clever design because the bones are very efficient at transferring loads, while the mechanism is precise at controlling

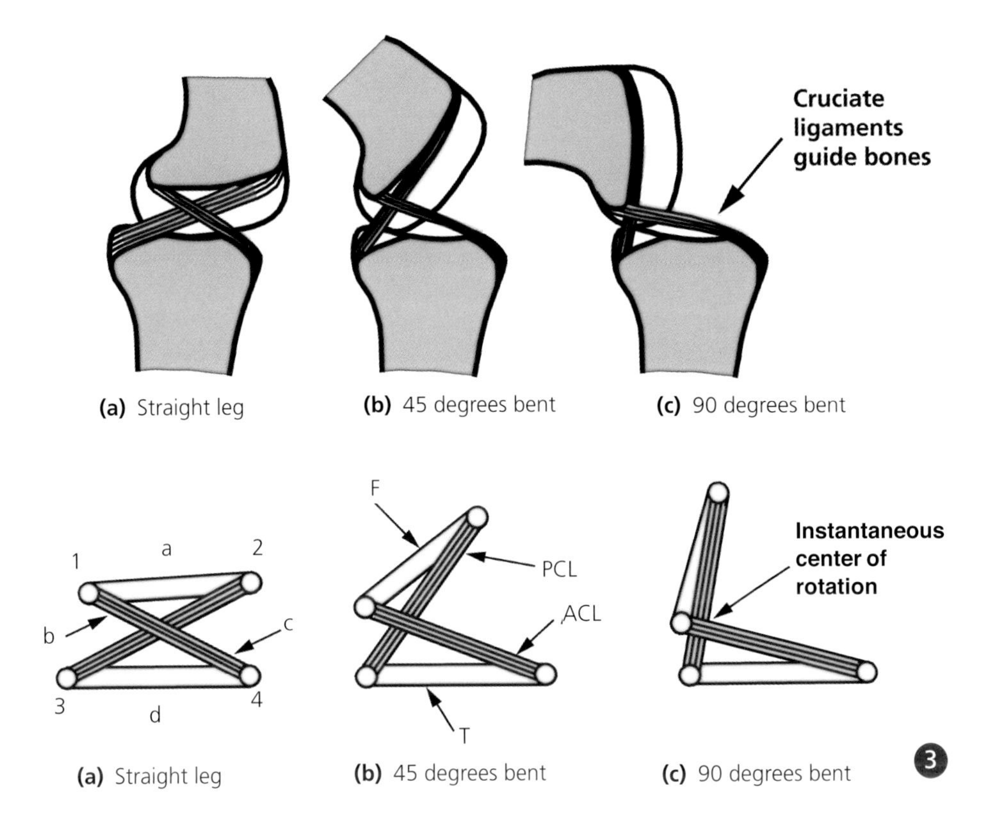

3 The two cruciate ligaments and two bones form a 4-bar mechanism in the knee joint

motion. One of the amazing design aspects of the knee is that the motion produced by the mechanism is exactly compatible with the rolling motion of the knee.

An important unique feature of the human knee is that it locks in the upright position. This makes it easy for humans to stand upright because it is not necessary to tense any of the muscles around the knee joint to maintain an upright posture. The human knee joint is such a brilliant design that engineers are producing bio-inspired robot joints based on it.[2]

Unique feet

Human feet have a unique arched structure as shown in figure 4. This makes it easy to stand because a person can feel the front (ball) and the back (heel) of the feet and can maintain balance by putting the center of gravity between the front and back.

2 A.C. Etoundi, S.C. Burgess, and R. Vaidyanathan, "A Bio-Inspired Condylar Hinge for Robotic Limbs," ASME J. *Mechanisms and Robotics,* Vol. 5, Issue 3 (2013).

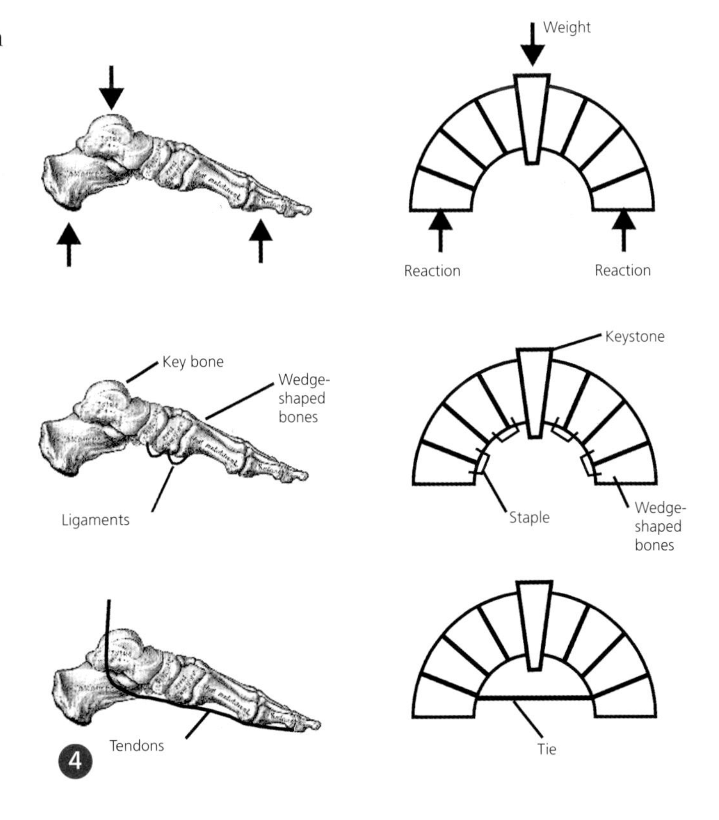

4 The human foot compared with a stone arch

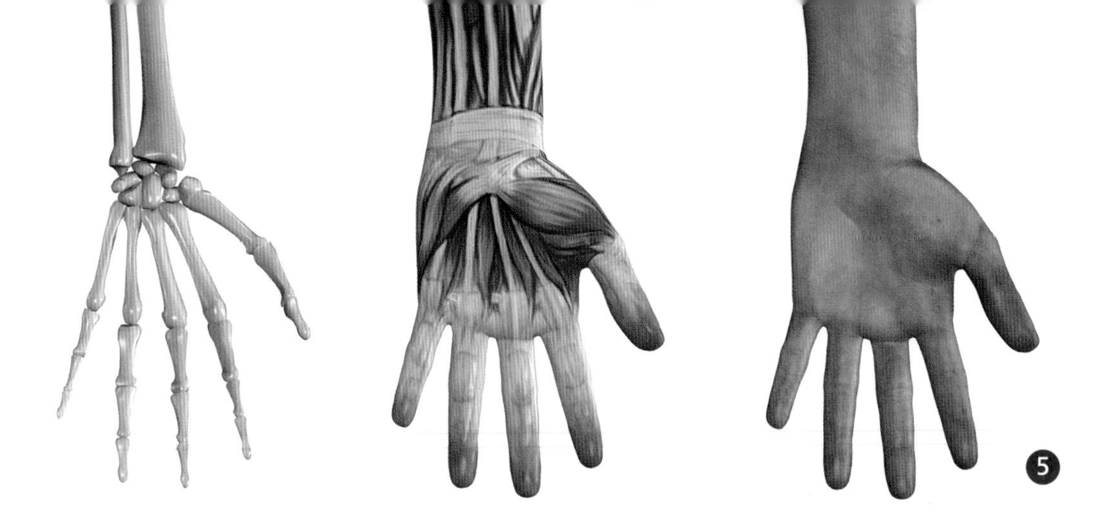

5 The anatomy of the human hand.

The human foot has 26 bones that fit together precisely to form a strong arch. There is even a wedge-shaped "key-bone" at the very top of the arch that stops the arch from collapsing. The key-bone is just like the wedge-shaped keystone in a man-made stone arch.

Humans are the only mammals that can run on two legs, because only humans can push off from the front of the feet and only humans have a forward-pointing big toe. Feet are well designed to give a lifetime of service. In a lifespan of 80 years, an active person walks over 100,000 miles, making over 100 million steps!

As far back as the 15th century, Leonardo da Vinci, who had a great knowledge of engineering, science, and anatomy, claimed, "The human foot is a masterpiece of engineering and a work of art."[3] Many great scientists and engineers down through the ages have come to the same conclusion.

Uniquely skillful hands

The human hand contains muscles and nerves that allow exceptionally fine movement of the fingers and thumb. Each hand is controlled by around 35 muscles, most of which are located in the arm in order to keep the hands slim. The muscles are connected to the hand via a set of slender tendons that are routed with extreme precision (figure 5). Engineers sometimes use cables to transfer a force over a distance, as with bicycle brake cables, but the best engineered cable system is very simple compared to the network of tendons in the human hand.

Human fingers have a unique full range of movement from a completely straight finger to a tightly curled finger. Another unique design feature of the human hand is an "opposable" thumb that can make face-to-face contact with the end of each finger. These pinch grips are the key to skillful tasks in areas such as writing, drawing, craftwork, cooking, and surgery.

The part of the brain responsible for muscular movements in the body is called the "motor cortex." About a quarter of the entire human motor cortex is devoted to controlling the muscles of the hands, showing that humans have the software as well as the hardware necessary for skillful hand movements.

According to evolution, human hands have evolved to throw spears and throw a punch. However, skillful hands are an example of where humans are designed for far more than survival. Hands are clearly designed for skillful and creative tasks such as playing musical instruments and creating works of art. In contrast, animals have no ability or desire to create works of art with their "hands."

Facial muscles for facial expressions

The human face has around 50 muscles, as shown in figure 6! About half of the muscles of the face are needed for physical tasks like eating, speaking, and closing the eyes. However, half of the facial muscles are dedicated to making facial expressions. For example, the muscles that go across the cheek bone and connect to the upper lip are the muscles that produce different kinds of smiles.

3 http://discoveringdavinci.tumblr.com/post/32222942704/the-human-foot-is-a-masterpiece-of-engineering.

To make facial expressions a person must learn specific combinations of muscle movements. The ability to move the right combination of muscles is learned mostly during early childhood when the whole muscular and nervous system is developing. Some expressions, like smiling, need only around four to six muscles to be activated. Other expressions, like frowning, can involve the use of up to 20 muscles.

Another reason why humans can make facial expressions is that the whites (sclera) of the eyes can be clearly seen when the eyes are open. The whites of the eyes sometimes emphasize certain facial expressions. Having whites of the eyes that are visible also makes it possible to see the direction of a person's gaze.

There are many different types of expressions that can be made, such as in smiling, disapproval, confusion, grief, anger, pain, surprise, and boredom (figure 7). By contrast, animals have very little ability or desire to express such complex emotions. The ability to make such facial expressions shows that humans are not animals, but emotional beings made in the image of God.

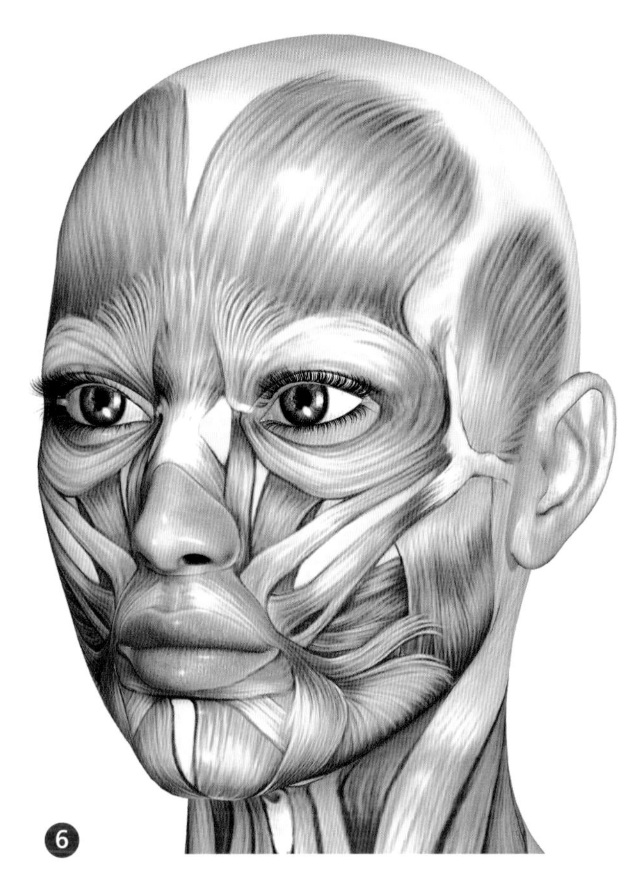

6 The muscles in the human face

7 Examples of human facial expressions

The Skin

Despite its simple appearance, human skin has an incredibly intricate design and performs many important functions for daily living.

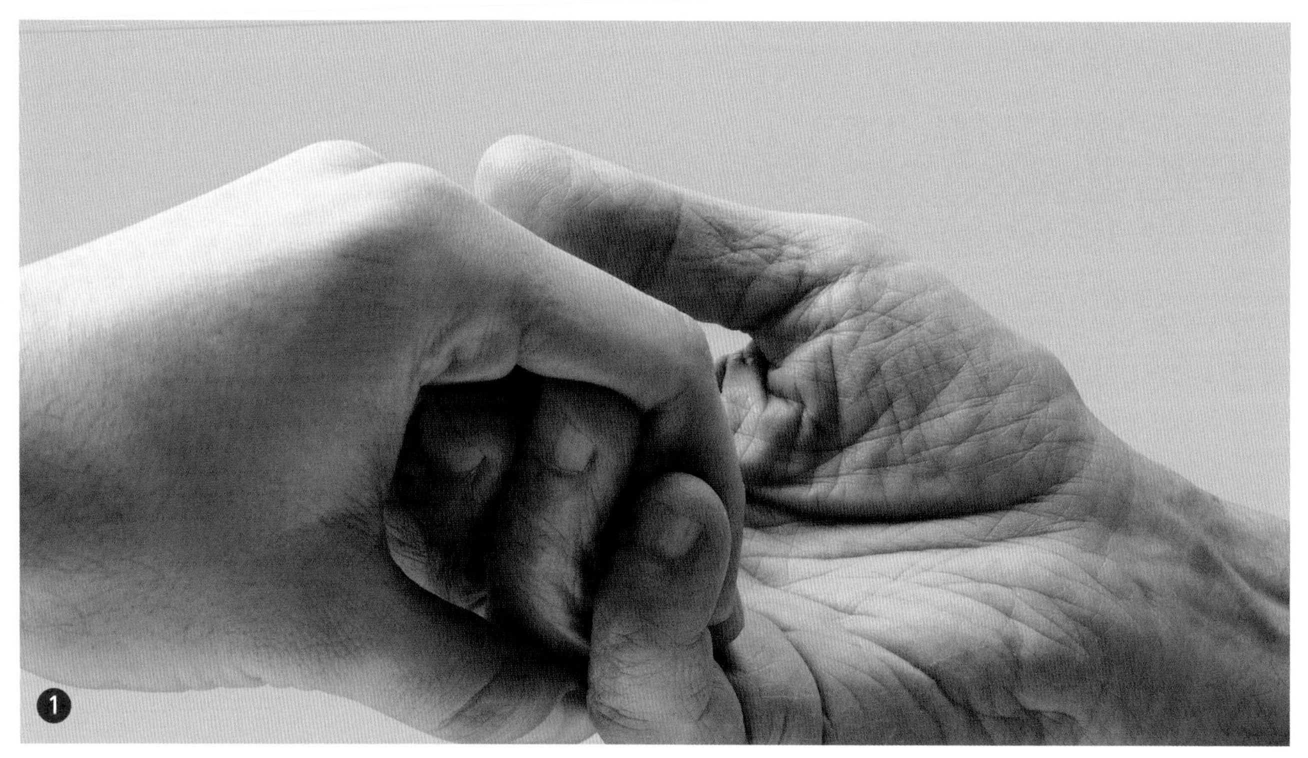

1 Hands have especially sensitive skin.

Skin is the largest organ in the body, covering an area of around 21.5 square feet(2 sq. m) in an adult. The thickness of skin varies from around 0.02 in (0.5 mm) for the eyelids to around 0.12 inches (3 mm) for the soles of the feet. As well as providing a tough physical barrier, skin provides protection from microorganisms and ultraviolet light. Skin plays an important role in temperature regulation through insulation and sweating. The fact that we need to wear clothes for insulation is a great advantage because we can adjust our clothing to suit our surroundings and activity. (It should also be noted that the primary purpose of clothing is to cover up nakedness following the Fall of mankind as described in Genesis 3.)

Skin prevents fluids from entering the body and, at the same time, regulates the amount of water leaving the body. Apart from the palms of the hands and the soles of the feet, skin has oil glands for keeping the skin smooth. These are usually located alongside hair follicles so that both the follicle and the gland can use the same opening in the skin. Skin even produces vitamin D from sunlight. Hands (figure 1) are particularly sensitive with many nerve endings connected to the skin.

A helpful feature of human skin is that everyone has a unique fingerprint which is very useful for crime detection.

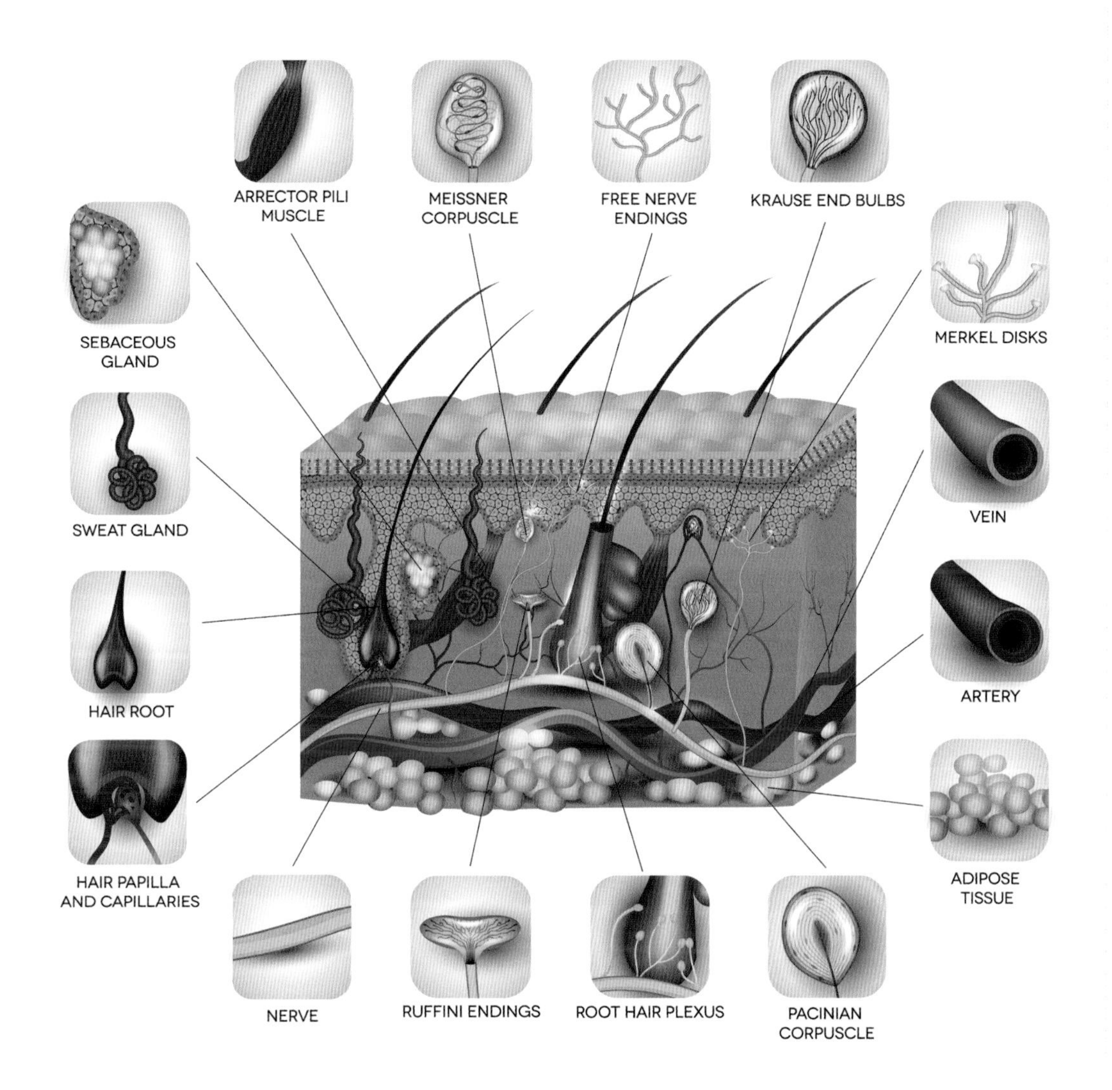

2 A simplified diagram of the structure of human skin

A tough, sensitive, and beautiful material

Figure 2 gives an idea of the amazingly intricate design of human skin. Skin contains hair follicles, sweat glands, blood vessels, heat sensors, touch sensors, pain sensors, oil glands, and nerves, all elegantly integrated into a multi-layered material. Each individual part has its own sophisticated design. Hair follicles have their own blood supply, dedicated muscle, and connection to the nervous system. Sweat glands are similarly well designed, with a long looping duct, having its own blood supply and a mechanism for extracting water from the blood.

Human skin consists of three layers as shown in figure 2: the outer layer, known as the epidermis, the middle layer, known as the dermis, and the inner layer, known as the hypodermis (sometimes called the subcutaneous layer). The epidermis is itself made up of a number of separate layers, each with different functions, and yet is only around 0.004 inches (0.1 mm) thick.

The hypodermis provides an anchor to the underlying tissues while still allowing flexibility so that the skin does not feel tight during body movements. This inner layer also provides insulation from hot and cold temperatures and protection from mechanical impact.

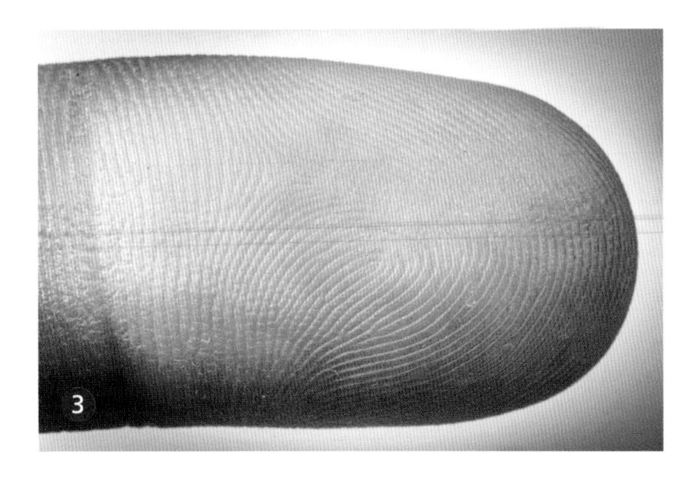

3/4 The fine touch sensitivity of human fingers makes them ideal for touch screens and braille.

Special sensors

There are four main types of sensor in the skin, enabling us to feel cold, heat, pain, and touch. There are also different touch sensors for the onset of pressure and for continued pressure. Fingertips (figure 3) are endowed with a high density of touch sensors. This enables us to carry out skillful tasks in areas such as craftwork, cooking, music, keyboards, touch screens, and braille (figure 4).

Different parts of the body also have different levels of sensitivity to touch and pain, with vulnerable areas like the face being more sensitive. A remarkable feature of the fingertips is that they are both touch-sensitive for delicate tasks and tough for constant use.

A notable aspect of human skin is its ability to cope with extreme loading and, at the same time, detect tiny loads. For example, it can feel a tickle from a feather and yet support and detect the weight of the whole body when running.

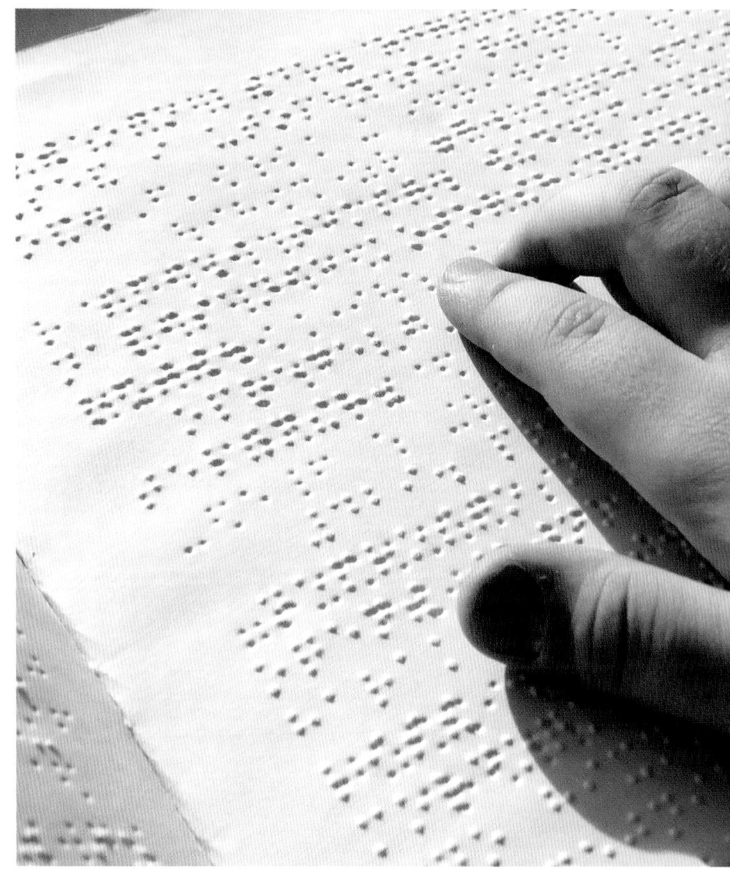

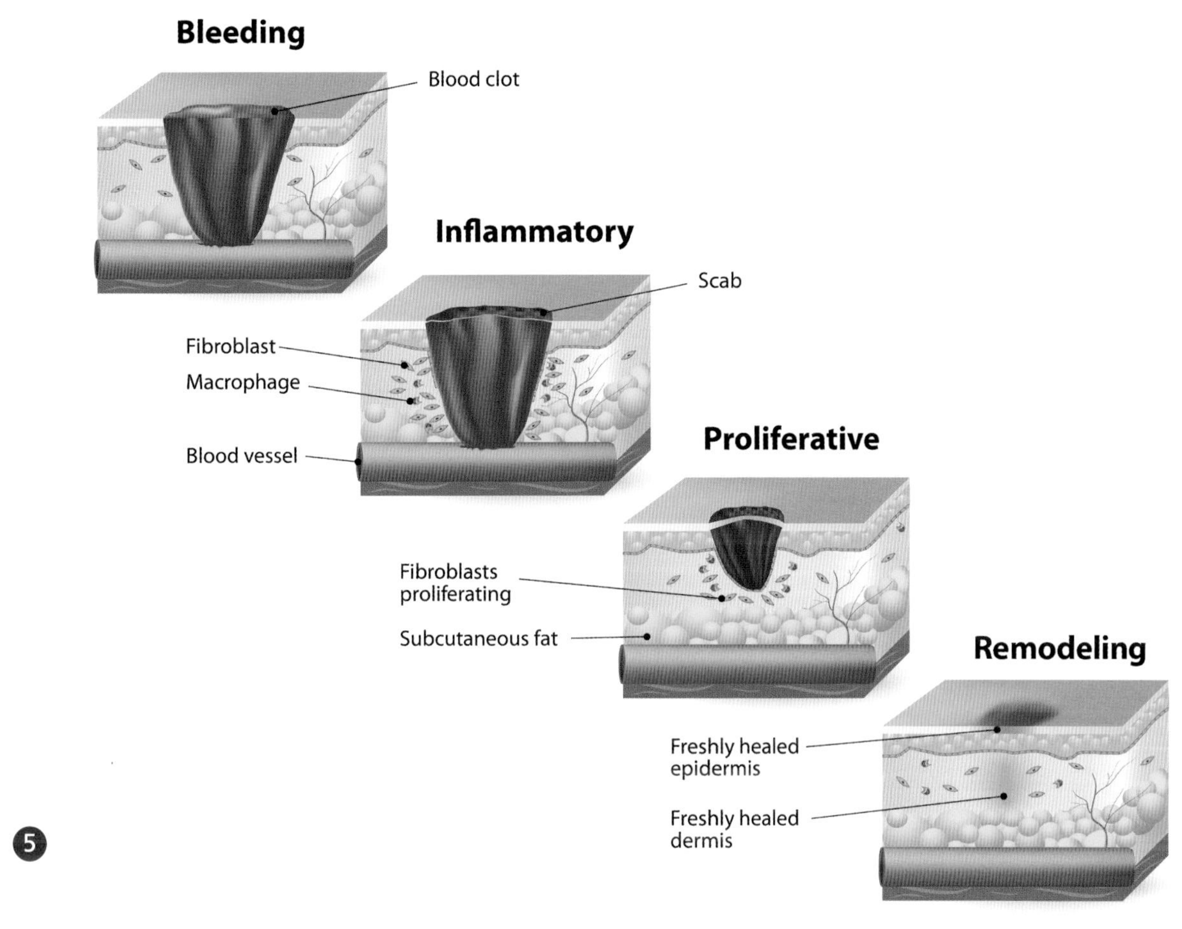

5 The healing process in skin **6** A knee scab

Ability to self-heal

Of all the properties of human skin, one of the most wonderful is its ability to maintain and repair itself (figure 5). New cells are constantly pushed to the surface so that it is always being replenished. This enables worn skin to be replaced and keeps the skin smooth and new-looking. The body can detect cell density in the skin and it uses this information to regulate the production of new cells. When cell density decreases due to skin wear or injury, cell division occurs to make new cells. After an injury like a cut, new cells gradually fill the gap and, once normal cell density is reached, cell division slows down to the normal rate. A scab stays on the skin until healing takes place (figure 6).

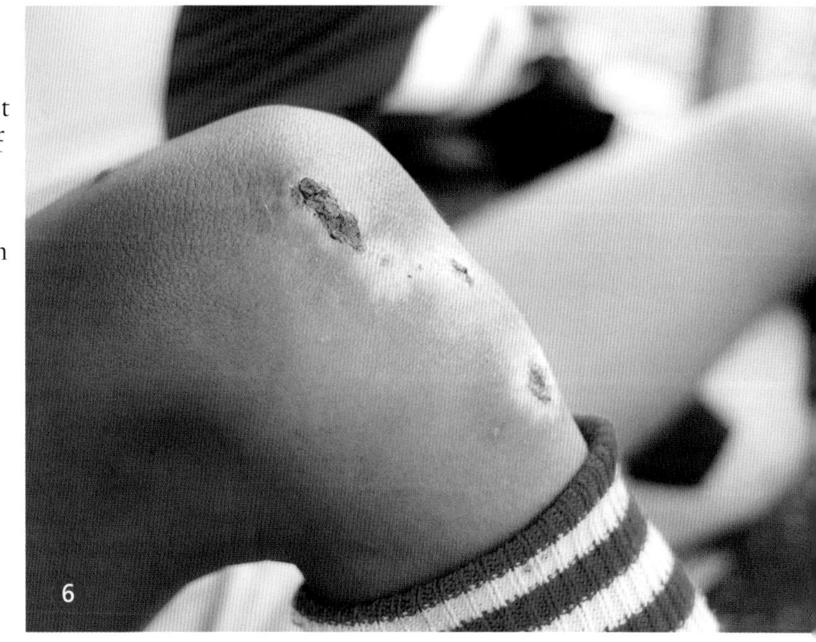

The Eye

Eyesight is something we tend to take for granted, yet it involves amazingly sophisticated optical design far beyond anything human designers can achieve.

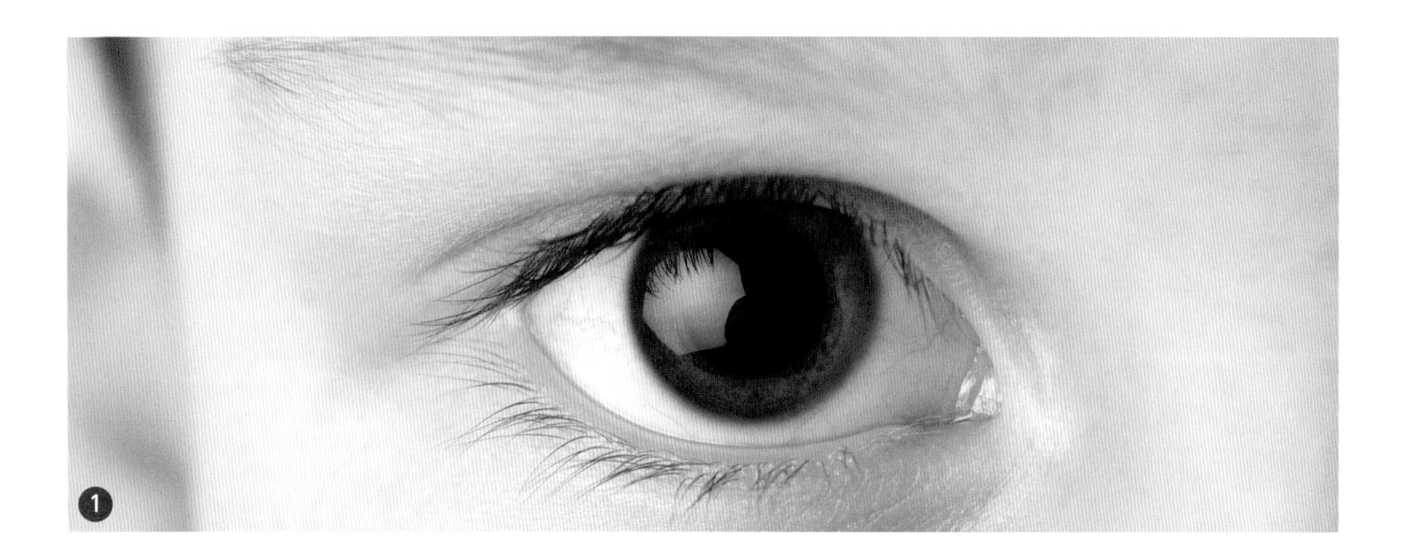

A living camera

The human eye looks simple from the outside (figure 1), but behind the beautiful exterior is a masterpiece of engineering design. The eye has similar components to those of an engineered camera. The cornea (figure 2) is like a lens focusing the light onto the retina. The lens located directly behind the pupil further focuses light like an autofocus camera lens. The iris is like the diaphragm of the camera, controlling the amount of light reaching the retina by automatically adjusting the pupil size.

The retina at the back of the eye acts like an electronic image sensor of a digital camera, converting optical images into electronic signals. The eyelid is like a camera cap and the tear duct provides a built-in cleaning fluid. The eye even has muscles that allow it to move like a motorized camera.

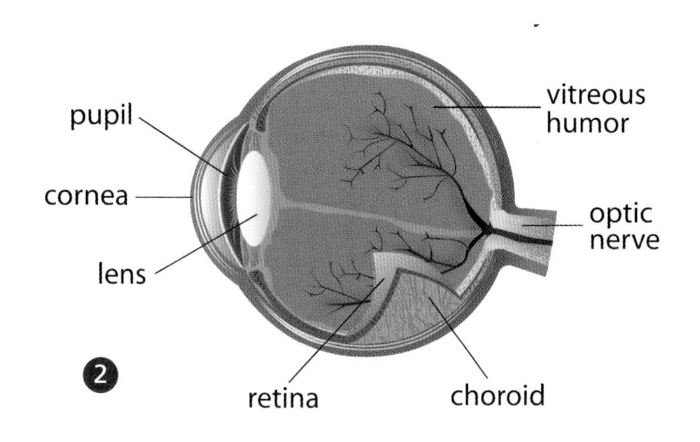

1 The human eye **2** Parts of the human eye

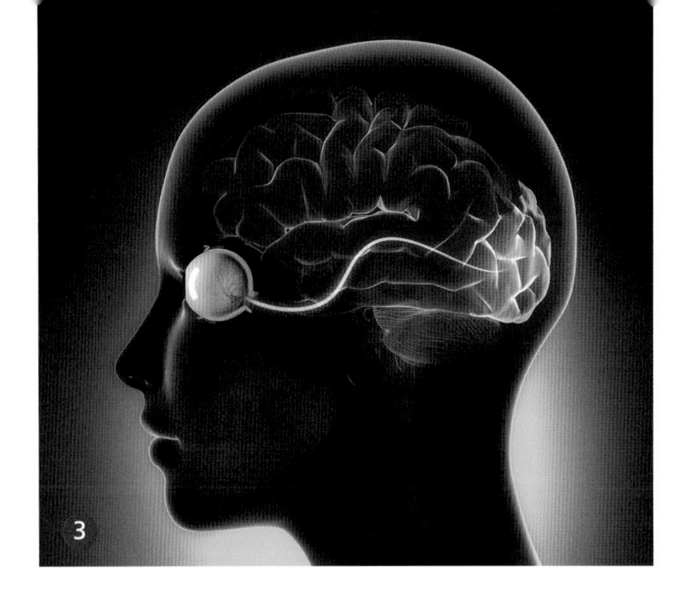

3 The optic nerve

The optic nerve is like a computer cable, transmitting the light signals to a part of the brain called the visual cortex where images are processed (figure 3). The recognition of a simple object like a cup is very straightforward, even for a young child, but it is actually a complex task involving the recognition of features such as depth, shading, and orientation. The human brain can correctly identify images in as little as 13 milliseconds.[4] In contrast, computerized robots often struggle to identify simple objects.

Brilliant color vision

Color vision in daylight is detected by cone cells. There are about six million cones in a human eye, resulting in a very high resolution of color images (figure 4).

At low levels of illumination, light is detected by rod cells in the retina. Rod cells cannot detect color, but are so sensitive that they can respond to a single photon of light (a photon being the smallest unit of light). When light is between dark and bright, both rods and cones provide signals.

Brilliant design, not bad design

There have been claims that the human eye contains bad design because the rod and cone cells are at the back of the retina, forcing the incoming light to travel through a mass of nerves in the front of the retina to reach the photoreceptors (figure 4).[5] But there are two problems with this claim. First, bad design is only bad if it leads to bad performance and this is not the case with the human eye which works perfectly well. Secondly, scientific research has revealed that when light reaches the retina it does not pass through a mass of nerves but gets funneled through special cells (called "Müller" cells) that act as precision optical fibers. These optical fibers not only funnel light to the photoreceptors but also improve the quality of the light by, for example, blocking scattered light.[6]

Another criticism of the human eye is that the optic nerve passes through the retina, creating a blind spot that has no photoreceptors. However, the blind spot does not reduce vision quality, because the other eye and the brain fill in the gap and the blind spot is located in a region used only for peripheral vision.

When the eye is considered as a complex functioning system, it becomes clear that it is not bad design but brilliant design. The discovery of optical fibers in the human eye is an example of how evolutionary thinking discourages good science because evolution assumes that the eye has a crude design and that there is no point looking for advanced features like Müller cells.

4 See http://news.mit.edu/2014/in-the-blink-of-an-eye-0116.

5 For example, Richard Dawkins, *The Blind Watchmaker* (1986) page 93.

6 Franze et al., "Müller Cells Are Living Optical Fibers in the Vertebrate Retina," *Proc. National Academy of Sciences* USA 104(20):8287–8292, May 15, 2007.

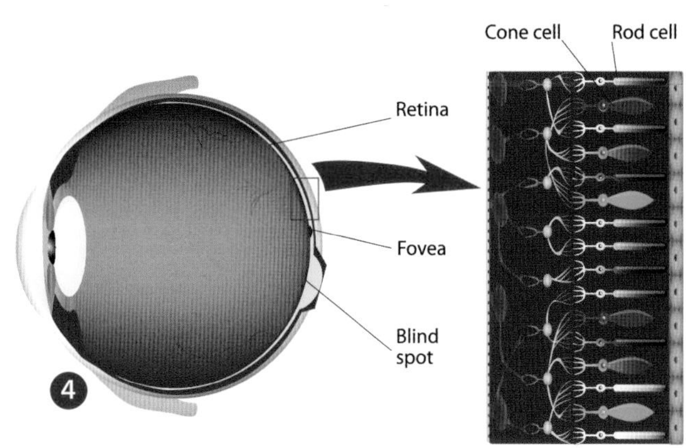

4 Photoreceptors in the retina

The Wonder of Hearing

The exquisite and irreducibly complex design in the human ear must surely point in the direction of a Creator.

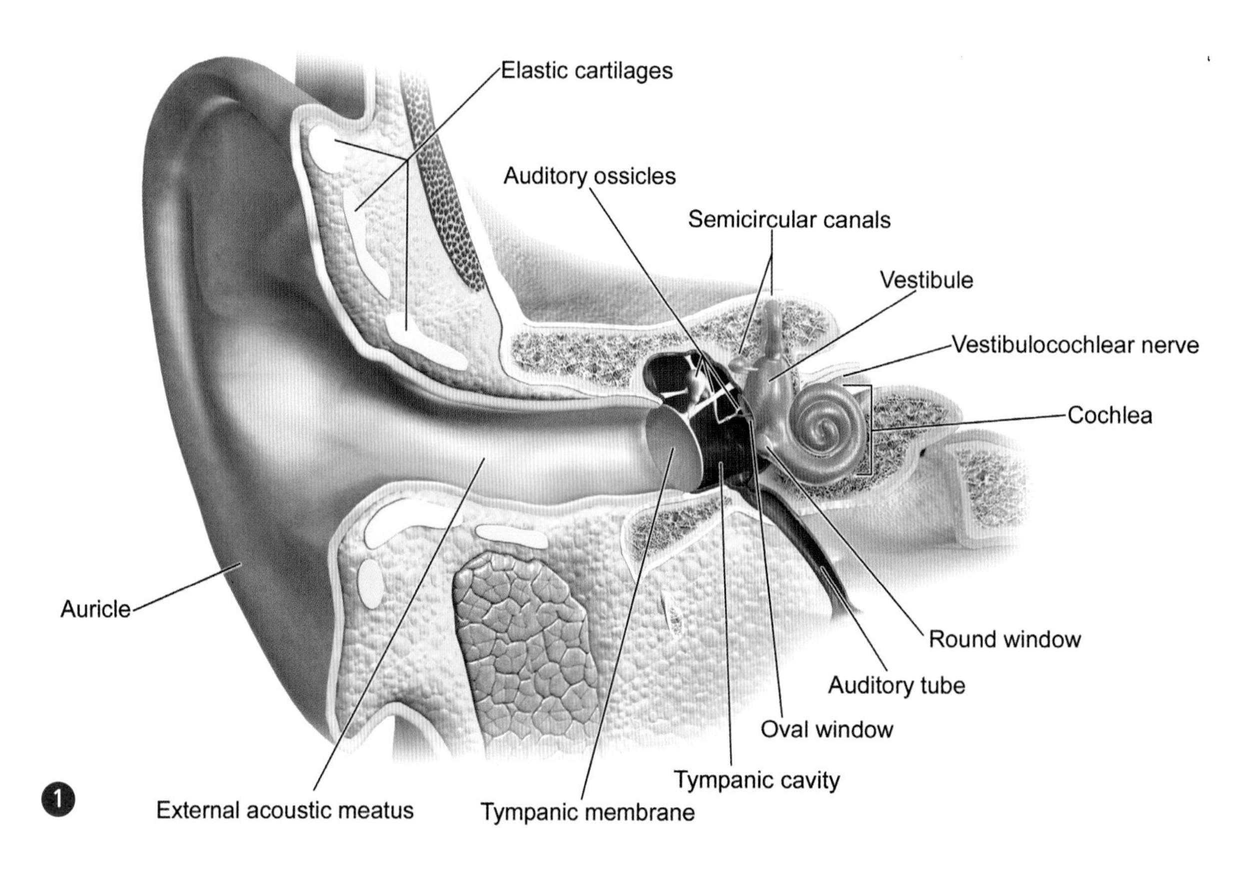

1 The anatomy of the ear

The way we hear

Sound is actually a pressure disturbance in the air such that small cyclical vibrations of pulsing pressure (called "acoustic waves") travel into the ear canal (figure 1) and then are transferred through tiny ossicle bones into the inner ear (figures 2/3). Each stage of this system is staggering in its complexity.

All mammals have a system which transfers acoustic waves through the eardrum and ossicle bones into the inner ear. However, there are large differences within the mammals. Certain length tubes vibrate at natural frequencies. The note from a flute will alter according to the change in its natural (resonant) frequency, which in turn is determined by where the musician presses open the keys. The ear canal in humans is about 0.8 inches (2 cms) long, while the ear canals in cats and dogs are of a different shape so that there is a bend in them. Thus, they have a vertical and horizontal ear canal which is designed for a different range of frequencies.

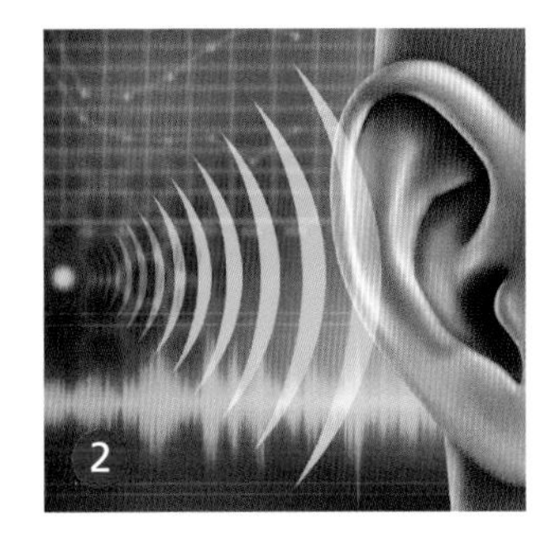

2 We hear sound as small cyclical pressure changes in the air which reach the ear drum inside the ear.

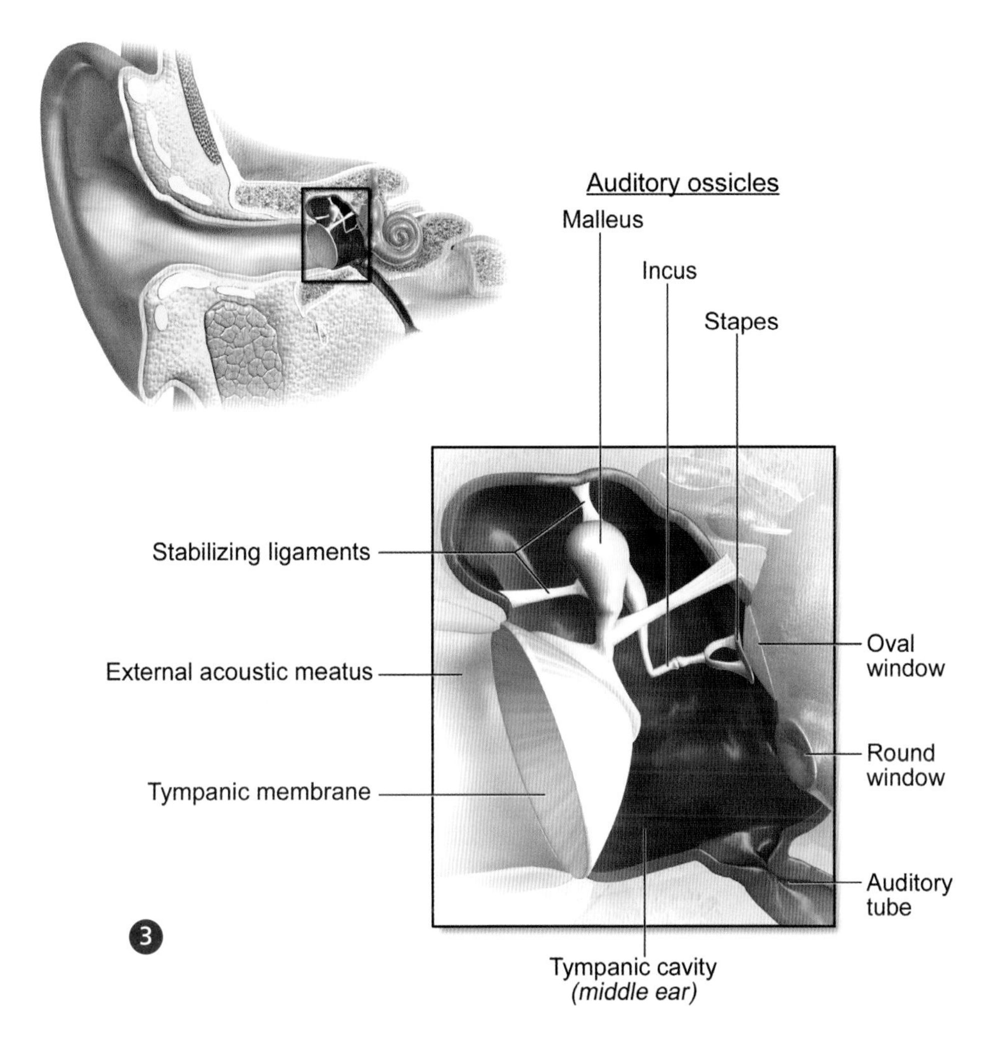

3 The ossicle bones

Humans hear over a wide range (9 octaves) from approx. 20 cycles per second (Hz) to nearly 20,000 Hz. Dogs hear from approximately 65Hz to 44,000Hz (again over 9 octaves but shifted compared to humans), while cats have one of the widest ranges of all: over 10 octaves from 55Hz to 77,000Hz.

Our ears are designed for low frequency as well as high frequency sound and resonate in the range of 4,000 Hz, which is right in the center of the human speech range. With the range of low and high frequency, our ears are especially able to appreciate many forms of music as well as speech. The higher frequency also enables us to appreciate bird song and other aspects of nature. Generally, the very young have the sharpest hearing. We begin to lose the upper 15–20KHz sometimes as early as in our 20s and 30s.

The middle ear — ossicle bones

The acoustic signal causes the eardrum to vibrate and push on the *malleus* (hammer) attached behind, which itself then pushes onto the *incus* (anvil) bone, which then moves the *stapes* (stirrup) horizontally (see figure 3). These bones are of the order of 0.2 inches (5mm) long, with the stapes smaller still, and all three can fit on a small coin with ease! These are the only bones in the body which do not grow. In adults, they are the same size as when we were babies. Those believing in evolution try to argue that upper and lower parts of the jawbones of a reptile moved to become the malleus and incus bones. But they quietly ignore the biggest hurdle to such a story, which is that the jaws of reptiles never stop growing, and after birth the ossicle bones in mammals never grow at all!

The ossicle bones amplify the signal. Each of the three are shaped specially so there is a lever mechanism such that the stapes (attached to a membrane called the oval window in the cochlea), moves approximately three times the distance traveled by the malleus. There is also a ten-fold smaller area being vibrated in the oval window compared to the tympanic membrane of the eardrum[1] so that the energy transfer involved is such that the system is almost 100% efficient.

Why the need to amplify the signal? This is because the signal is now going to pass into a liquid medium in the inner ear!

The inner ear — cochlea

Liquid is a barrier to sound, and the stapes acts like a pump on the oval window membrane and, cleverly, the membrane of the round window (see figures 3 and 4) expands to compensate for the movement of the liquid inside the cochlea.

Unwinding the cochlea (figure 5), we can see an ingenious basilar membrane which tapers for higher frequencies inside the cochlea, rather like a xylophone, so that whatever combined frequencies come in from the oval window vibration are immediately split up into their component frequencies. This is effectively an instantaneous frequency analyzer, so that whatever signal is coming in, composed of many frequencies, immediately causes *different* parts of the basilar membrane to vibrate.

The final part of the hearing system is achieved by the organ of Corti (see detail of figure 4) running along on top of the basilar membrane. This has tiny little hairs *(stereocilia)* on it (figure 6) which send an electrical signal according to each frequency excited by the incoming signal. It is astonishing that each cilia (0.00025 mm thick — this is less than 1/70th of the thickness of a human hair!) when disturbed by the tectorial membrane (which touches the cilia above) has a little trapdoor at its side that opens with a spring attached (see figure 6) to an adjacent cilia! This allows electrical ions in the electrically charged fluid within that part of the cochlea, to then excite

1 https://commons.wikimedia.org/wiki/File:Areal_Ratio_of_tympanic_membrane.png. Accessed Sept 2016.

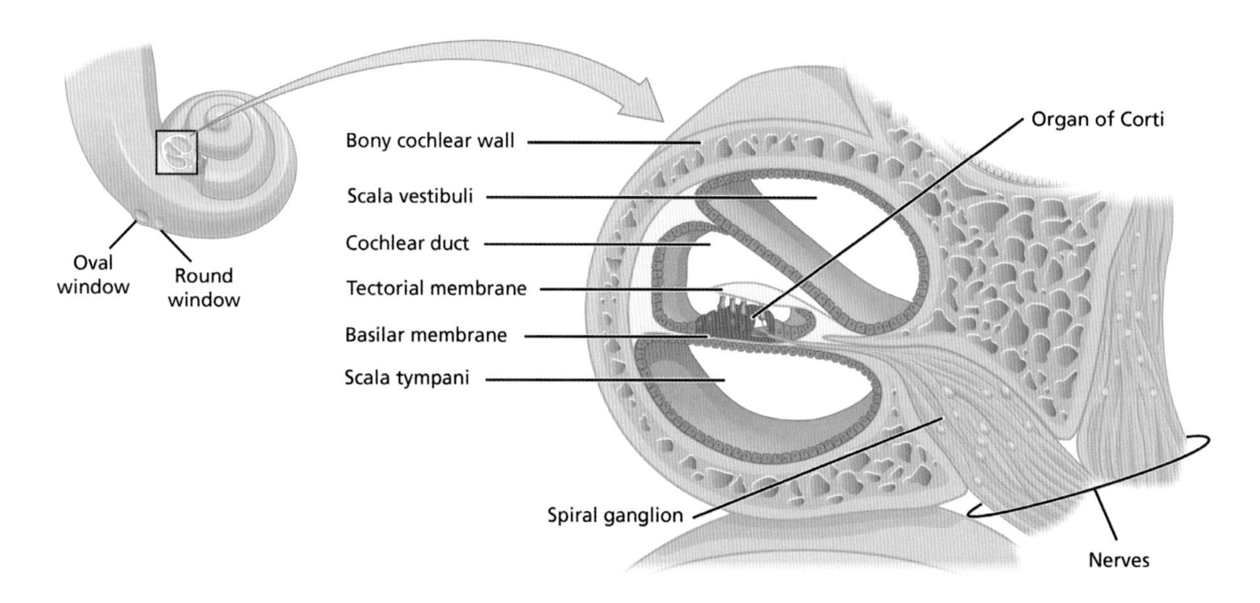

4 The cochlea and organ of Corti

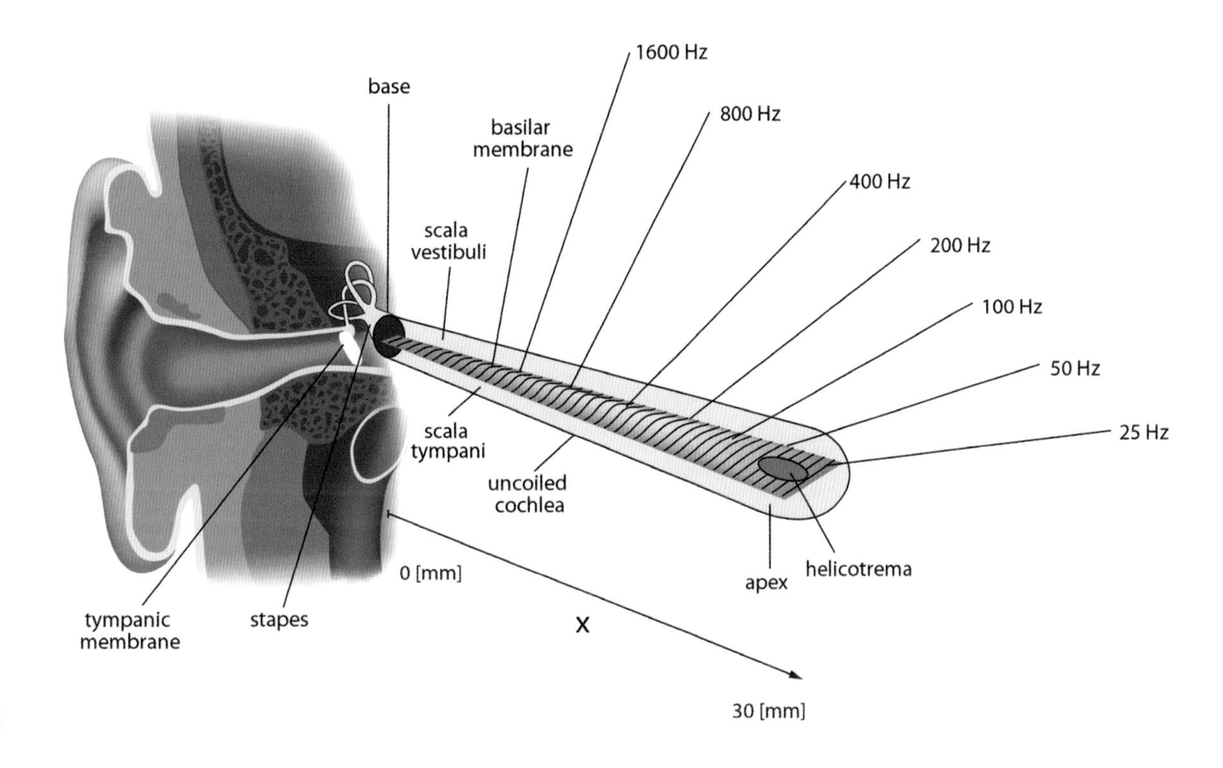

5 The Cochlea unwound with basilar membrane visible — rather like a xylophone. *(Courtesy of A. Kern, C. Heid, W.H. Steeb, N. Stoop, R. Stoop, Wikimedia)*

6 The cilia have a trap door operated by a spring attached to an adjacent cilia. The charged ions then move down the cilia and excite the nerves to the brain. A very similar system is used for the three semi-circular canals which control balance in the inner ear.

ganglion nerves to send the signal to different parts of the brain, depending on whether it is music or speech. For low frequencies there is about one nerve for each change in Hz. In the upper range it is about 2–3 Hz per nerve ending.

Such a system involving air vibrations, mechanical, chemical, and electrical engineering confirms the intelligent design of the ear.

"The hearing ear, and the seeing eye, the LORD has made them both" (Proverbs 20:12).

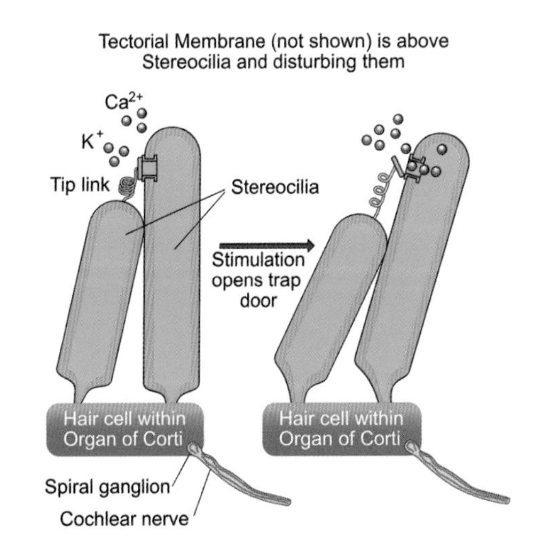

12 A Changed World — Rock Layers

The layering of rocks across the world provides evidence for a catastrophic global Flood in the more recent past.

The story of the rocks — massive layering and upheavals

Rocks laid down by water are called sedimentary — approximately 70% of the land surface is laid down by water. This is a surprising statistic for the evolutionary scenario which believes that these rocks have gradually formed over millions of years of water deposit and wind erosion. However, it is entirely to be expected if sedimentary rocks are primarily a result of the worldwide Flood. We saw on page 111 that, amazingly, even Mars has sedimentary rocks and most admit that water flowed there extensively in the past.

The depth of sedimentary rock on our planet in places is spectacular. A most impressive site displaying the sheer magnitude of layering is the Grand Canyon (figure 1 and page 200, figure 5) where vertical sides tower over the Colorado River one mile below. The canyon is 277 miles (446 km) long, up to 18 miles (29 km) wide, and reaches a depth of over 1 mile (1.6 km). Geologists connect the layers right across the globe to construct what is called the geological column. Work begun by William Smith[1] strongly indicated the interconnected picture of national (English) deposits, and from this the modern study of geology led to the understanding of the global interconnected deposits. But are these layers slowly deposited? There is much evidence that they were formed rapidly. For example:

1 The Grand Canyon — an example of massive horizontal layering. *(Photo by Andy McIntosh)*

1. The vast parallel lines in the layers with very little evidence of weathering above any individual stratum (see for instance figure 1) suggests no time between deposits.

1 W. Smith, "A delineation of the strata of England and Wales: With part of Scotland; exhibiting the collieries and mines, the marshes and fen land originally overflowed by the sea, and the varieties of soil according to the variations in the substrata, illustrated by the most descriptive names" (London: John Cary, 1815.) See also J. Phillips, *Memoirs of William Smith*, 1844, reprinted by Bath Royal Literary and Scientific Institution, 2003.

2. Unconformities in the layers. This term is used when layers supposed by some to be laid down slowly over millions of years, are in fact missing. A famous example is the Great Unconformity in the Grand Canyon where Tapeats sandstone, supposedly 550 millions of years old, sits on top of Precambrian rock which is dated at 1,700 million years old. Thus, there is a gap of 1,150 million years which is represented by a single flat contact line showing no erosion which one would expect if the rocks were laid down slowly. This is no issue if this is the action of vast amounts of water laying down sediments over continents, and not always exactly the same at every location.

3. Often the layers are twisted and contorted, indicating that they have been bent while soft. Figures 2 and 3 are very striking examples of folding consistent only with fast movement of water and twisting of soft sediments.

There are even cases where the layers have been folded, followed by a vast sweeping action which has then sheared off the tops of these folds. The Matterhorn in the Swiss Alps and Mount Everest are both examples where the reverse of the layers of the Geological Column takes place — supposedly older rock sits on top of younger.

4. Often the layers are twisted and turned such that large amounts of sediment are almost on end at nearly 90 degrees. There are many cases of this in the western part of the British Isles in Devon, Snowdonia, and the western Highlands of Scotland. Other mountain ranges across the world show tilting of thousands of feet of layers in the Alps, the Rocky Mountains, and Canada, Northern Iran (figure 4) is a dramatic example, and Uluru (Ayers Rock) in Australia.

All these examples indicate that the sedimentary layers themselves were catastrophically deposited.[2]

2 A recumbent fold in the Coconino at Lizard Head, near Sedona, Arizona — no fracturing. *(Photo by John Whitmore)*

3 A folded sedimentary layer in lower Ugab valley in Namibia

4 A cutting along Route 59 in Northern Iran on Chalus Road showing sedimentary layers tilted on end

2 Much more detail of these principles is in books such as H. Morris and J. Whitcomb, *The Genesis Flood* (P&R Publishing, 2008) and J. Whitcomb, *The World that Perished* (Baker, 1988), A. Snelling, *Earth's Catastrophic Past* (ICR, 2009), P. Garner, *The New Creationism* (Evangelical Press, 2009).

Some have recognized this principle though still keeping to an evolutionary framework. See D. Ager, *The New Catastrophism* (Cambridge University Press, 1993).

"Living" Fossils

Fossils certainly tell a story, but not of slowly emerging life — rather one of rapid death and catastrophic burial. Many of them are essentially the same as creatures today.

1 Mass mortality of fish — *Knightia eocaena* — Green River Formation, Fossil Butte National Monument, Wyoming *(U.S. National Park Service)*

2 Fossil Fish — *Diplomystus dentatus* with *Knightia* in its mouth *(Photo U.S. National Park Service)*

Though rare in the rocks, fossils are found all over the world, and almost invariably in sedimentary rock; there are occasional exceptions where volcanic rock has encased a creature or vegetation.

Not only the rock but the fossils also show catastrophe as shown in the Green River Formation of Wyoming (figure 1), where exquisite detail can be seen on each of the fish including scale markings. Perhaps the best example is figure 2 where a fish is buried in the process of eating another fish! This is strong evidence that these creatures were buried rapidly — a snapshot in time at the moment of eating.

Many fossils are essentially the same as creatures today. We commented on this when considering fossils of butterflies and dragonflies in chapter 5. The fish in figure 1 are similar to herring today. Giraffes,[3] crocodiles,[4] octopuses,[5] and many more species are found preserved in the rocks. The major difference is that some fossils are larger than their modern counterparts. Even the coelacanth fish — once thought to be extinct and only known from Devonian rock supposedly 380 million years old — were recognized in 1938 to be the same as fish which had been caught for years in deep water off the coast of Madagascar.[6]

3 E.R. Lankester, *Extinct Animals* (London: Archibald Constable & Co. Ltd., 1905); http://www.tolweb.org/Giraffoidea/50877. Accessed July 2016.

4 http://www.nationalgeographic.com/supercroc/index.html and http://ngm.nationalgeographic.com/2008/09/supercroc/sereno-text. Accessed July 2016.

5 D. Fuchs, G. Bracchi, and R. Weis, "New Octopods (cephalopoda: coloeoidea) from the late Cretaceous (upper cenomanian) of Ha^kel and Ha^Djoula, Lebanon," *Palaeontology*, 52(1) 2009, p. 65–81.

6 T. Clarey and J. Tomkins, "Coelacanths — Evolutionists Still Fishing in Shallow Water," *ICR Creation Science Update*, April 29, 2013; see also K. Ham, "The Fish That Forgot to Evolve," Answers in Genesis, Nov 2015; https://answersingenesis.org/blogs/ken-ham/2015/11/04/fish-forgot-to-evolve/. Accessed July 2016.

Coelacanths have since been found in higher Cretaceous layers[7] and these are essentially no different from their Devonian counterparts (supposedly 280 million years earlier), and are similar to the modern ones alive in the Indian Ocean today. There has been adaptation, but no evolution into new creatures.

The fossil record has many examples of creatures we still have today called "living fossils."[8] Figure 3 shows a prawn not too dissimilar to those we enjoy in our prawn cocktail! Figure 4 shows a silky lacewing psychopsid; this is very similar to the psychopsids (lacewings) today. Notice the pigment squeezed into the rock from catastrophic burial.

In the plant world too, there are many examples where a fossil looks distinctly like plants that we still have now. Gingko trees, supposed by many to be ancient, are in fact exactly the same in the fossils as those growing today.[9] Figure 5 is an example of a delicate fern formation which has the same structure as a modern fern. How are fossil plants so exquisitely preserved? These fossils are made by minerals (primarily silica compounds) chemically reacting with the original living material to produce hard crystalline material in its place.[10] In the case of plant fossils, this reaction has to be fast, since, after being separated from its roots, the plant leaf would wither in air in a matter of hours, and in water it would quickly lose its robust leaf form. Everything is indicating that the burial happened quickly. In fact, there is rarely any indication of natural withering at all in plant fossils.

7 J. Graf, "A New Early Cretaceous Coelacanth from Texas," *Historical Biology* 24(4), 441–452, 2010.

8 C. Werner, *Evolution: The Grand Experiment* (Green Forest, AR:New Leaf Press, 2007), Vol.1 2007, Vol. 2 2008. In this book Carl Werner shows many examples of "living" fossils. See also AiG article on "living" fossils at https://answersingenesis.org/fossils/living-fossils/. Accessed July 2016.

9 D.L. Royer, L.J. Hickey, and S.L. Wing, "Ecological Conservatism in the 'Living Fossil' Ginkgo," *Paleobiology*, 29(1):84–104, 2003.

10 H. Piepenbrink, "Examples of Chemical Changes During Fossilization," *Applied Geochemistry* 4(3), 273–280, 1989. See also http://www.fossilmuseum.net/fossilrecord/fossilization/fossilization.htm for a good summary of fossilization processes.

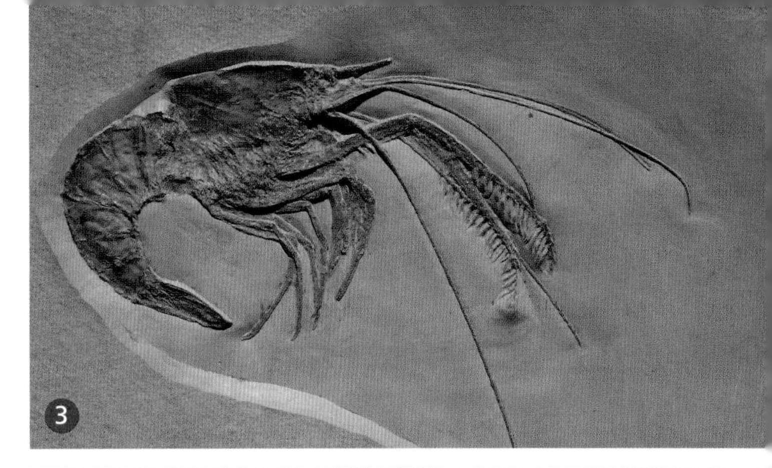

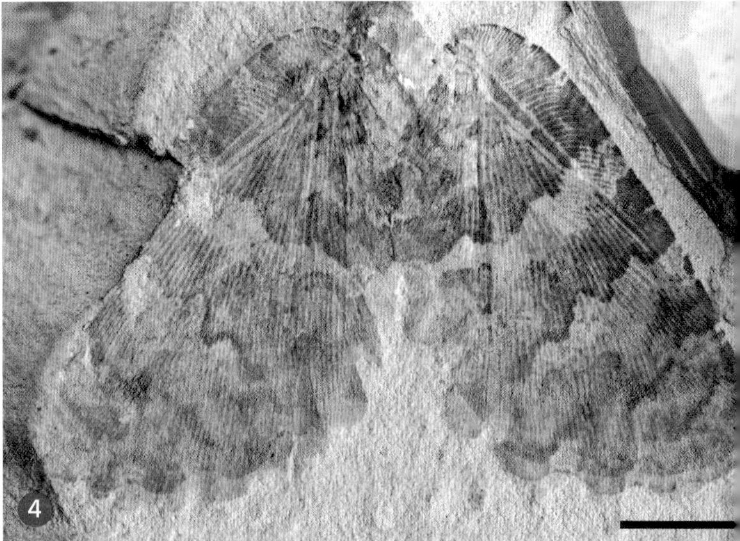

3 Fossil prawn — *Aeger elegans* from Cretaceous rock at Solnhofen, Germany

4 Silky lacewing Psychopsid Fossil — *Undulopsychopsis alexi*

5 Fossil fern from St Claire, Pennsylvanian Shale (late Carboniferous) *(Photo by Andy McIntosh)*

Extinct Creatures

Were extinct creatures evolving while the ones we have today stayed the same?

1 Phacopida trilobite from North Africa. *(Photo by Andy McIntosh)*

2 Phacopida trilobite compound eye with schizochroal lenses — that is, each lens has an individual cornea *(Photo courtesy of John Mackay)*

In this section we consider some examples of types of creatures that some claim were ancient ancestors to the rise of modern life today: trilobites and ammonites.

Trilobites

According to a consensus among evolutionists, trilobites are some of the "oldest" organisms known of the fossilized past. They were evidently sea creatures that walked on the sea floor. Phacopida trilobites are shown in figures 1 and 2. There is a large variety of trilobites and they are found in Cambrian through Devonian and Carboniferous up to Permian strata.[11] Thus, traditional dating for trilobite fossils, which are numerous among animal fossils, would be 250 million years to 550 million years ago. A remarkable fact is that precursors to these creatures have never been found. Arthropods (many-legged creatures) appear abruptly in the Cambrian strata, and do not show evidence of development; rather there is a large variety of trilobite fossils in these layers which strongly suggests that all these varieties always existed alongside each other.

What is fascinating also is that the compound eyes of trilobites are not made of protein but of calcite, which is a mineral! And the problem with natural crystals of calcite ($CaCO_3$) is that they give double images, (see figure 3). The reason for this is that all the light is polarized into two planes, and one plane of light is bent at one angle and the other at a different angle. Calcite is what we call a birefringent material. There is one, and only one, line of sight which will bring the two images together. This is called the optical axis, and the center line of any lens built from the calcite must be lined up to that optical axis. This is not an obvious line. It is frankly stunning that each of the hundreds of lenses in trilobites (and in some species it is thousands of lenses) is lined up exactly to the

11 C. Barnard, A.C. McIntosh, and S. Taylor, "Origins — Examining the Evidence," *Truth in Science*, 2011; see p. 83–86, "Trilobite Eyes: Out of the blue?" where the detail of the design features of the calcite lenses is described.

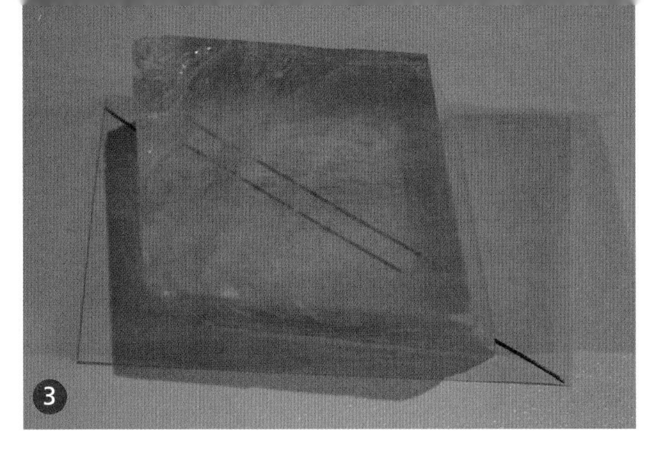

3 The double image occurring in a natural crystal of calcite (calcium carbonate, $CaCO_3$) *(Photo by Andy McIntosh)*

4 Ammonites are prolific in the Jurassic rocks of NE England — here many are all buried together in rock from Sandsend near Whitby, North Yorkshire. *(Photo by Andy McIntosh)*

optical axis — brilliant optical engineering in such a "primitive" creature!

Furthermore, there is a correction of the lenses for chromatic aberration — i.e., to avoid false color effects which one always gets even in glass lenses, due to red light being bent at a different angle to blue light. It was discovered by Descartes (1596–1650) and Huygens (1629–1695) that there are certain optimal shapes that remove spherical aberration from images and keep the image in sharp focus. It is quite astonishing that the shape of the trilobite lens is in accordance with those very shapes[12] discovered in the 17th century!

There is more. In some trilobites, extra intralensar material is also attached to the main lens that further sharpens the image. The upper part of this doublet lens is particularly important for seeing in water, since the additional material has a different refractive index and enables the sharp focus to be maintained over a long depth of field.[13] There is no doubt that trilobites were not simple creatures, and there is no evidence of evolutionary precursors. This is, in fact, brilliant optical engineering!

Ammonites

Ammonites, though extinct now, are similar to modern nautiloids and are found in large numbers in the fossil-bearing rock — usually in Jurassic rock such as the example shown in figures 4 and 5 from northeast England. They added sections as the animal grew inside its shell so that there are examples where the ammonites are 3 or 6 feet in diameter. Again, there are no precursors in the rocks. They are littered across the globe and even found in deserts in Israel (see figure 6) and in the foothills of the Himalayas. Their presence in the rocks speaks clearly of the Flood and its catastrophic consequences.

12 E.N.K. Clarkson and R. Levi-Setti, "Trilobite Eyes and the Optics of Descartes and Huygens," *Nature* 254 (1975): 663–667.

13 R. Levi-Setti, 1993. *Trilobites: A Photographic Atlas* (Chicago, IL: University of Chicago Press, 1993), second edition, p. 33.

5 Single ammonite with mirror impression from Jurassic rock near Whitby, North Yorkshire *(Photo by Andy McIntosh)*

6 Ammonites large and small buried in almost vertical sedimentary Cretaceous rock at Har Hanegev Nature Reserve, south of Beersheva in Southern Israel *(Photo by Andy McIntosh with friend Lior Regev giving sense of scale)*

Dinosaurs

Dinosaurs are a great favorite with many and conjure up ideas of ancient reptiles roaming the planet long before people were around — but is this perception true?

1 Fossil of a *Giganotosaurus* (Fernbank Natural History Museum Atlanta)

2 Fossilized skin AMNH 5427 of the horned dinosaur *Centrosaurus apertus* (Monoclinius nasicornus) found in Cretaceous layers near Steveville, Alberta, Canada. Plate 18 of ref. 18

Bones and footprints

There is no doubt that these creatures existed (figure 1). The issue is, when? The word "dinosaur" was coined by Richard Owen who was the first curator of the British Museum in London. It means "terrible lizard." Fossil footprints and remains of dinosaurs have been found on every continent and in numerous locations,[14] including Antarctica.[15] Most consist of bones and teeth because they are resistant to decay, but sometimes soft parts are discovered and there are rare occasions when not only an imprint[16] (figure 2) but actual dinosaur skin itself has been found.[17] This, of course, immediately raises the question concerning the age of such finds when conventional dating places these dinosaurs at least

14 G. Barnard, A.C. McIntosh, and S. Taylor, "Origins — Examining the Evidence," *Truth in Science*, 2011, see p. 87–95, "From Dinosaurs to Birds?"

15 http://www.dailymail.co.uk/sciencetech/article-3577625/Huge-trove-dinosaur-fossils-Antarctica-71-million-year-old-bones-reveal-new-clues-went-extinct.html. Accessed August 2016.

16 Dinosaur skin impressions! North Carolina Museum of Natural Sciences, July 9, 2012. Duck-billed dinosaur from Hell Creek formation, Montana; https://naturalsciencesresearch.wordpress.com/2012/07/09/439/. Accessed August 2016.
B. Brown, "A Complete Skeleton of the Horned Dinosaur Monoclonius, and Description of a Second Skeleton Showing Skin Impressions," *The Bulletin of the American Museum of Natural History*, Vol 37, Article 10, 281–306, 1917 – plate 18 is shown in figure 2.

17 "Test of Ancient Skin Sample Will Reveal Final Answer," *Nature World News*, natureworldnews.com, April 29, 2013; http://www.natureworldnews.com/articles/1649/20130429/what-color-dinosaurs-test-ancient-skin-sample-will-reveal-final.htm. Accessed August 2016.

3 *Tyrannosaurus Rex* skeleton at American Museum of Natural History *(Photo by J.M. Lujit. Wikimedia)*

65 million years old. Many trackways of dinosaurs have been discovered;[18] for example, those in Enciso, La Rioja, in the north of Spain. Even dinosaur eggs, embryos, and hatchlings are preserved in some locations.[19]

The analysis of the fossilized bones of dinosaurs found in the rocks shows that, based on differences in the structure of the pelvis, there are two different types of dinosaur: "lizard-hipped" (saurischians) and "bird-hipped" (ornithiscian) dinosaurs. There is great variety within these two groups of dinosaurs. The "lizard-hipped" dinosaurs include the bipedal theropods (assumed from their teeth to be flesh-eating dinosaurs) and the sauropodomorphs (the large long-necked plant-eaters). The "bird-hipped" dinosaurs also include many different types. There are the ornithopods (two-legged plant eaters), the pachycephalosaurs (the bone-headed dinosaurs, such as the extraordinary duck-billed hadrosaurs), and the ceratopsians (the horn-faced dinosaurs). The stegosaurs (with plates and spines) and the ankylosaurs (the armored dinosaurs) are also within this bird-hipped group.[15] There is also a big variation in size. The smallest dinosaur was likely to have been *Compsognathus* — a therapod which, from the two skeletons found, was approximately 3 feet (one meter) long.[20] The largest dinosaur fossilized skeleton ever found (in 2014[21]) is the sauropod *Argentinosaurus huinculensis* which was 130 feet (39.7 m) long, and it is estimated that it weighed 94.9 tons(96.4 metric tons)! [A metric ton (tonne) = 1,000 kg = 2,014.6 lbs; imperial ton = 2,240 lbs.] One of the largest tetrapods was *Giganotosaurus* (figure 1) estimated to be approximately 47 feet (14.3 m) long which is approximately 4 feet (1.22 m) longer than *Tyrannosaurus Rex* (figure 3). *Giganotosaurus* weighed between 7.9 and 9.8 imperial tons (8–10 metric tons), and had a skull which was itself as large as a man at about 6 feet (1.83 m) long.[22]

18 Dinosaur trackway at Enciso, La Rioja, Spain, https://en.wikipedia.org/wiki/Enciso,_La_Rioja. Accessed August 2016.

19 J.A. Wilson, D.M. Mohabey, S.E. Peters, and J.J. Head, "Predation upon Hatchling Dinosaurs by a New Snake from the Late Cretaceous of India," *Plos Biology*, Volume 8(3), March 2010; https://journals.plos.org/plosbiology/article?id=10.1371/journal.pbio.1000322.

20 https://en.wikipedia.org/wiki/Compsognathus. Accessed August 2016.

21 J. Morgan, "'Biggest Dinosaur Ever' Discovered," BBC News Science and Environment, May 17, 2014; http://www.bbc.co.uk/news/science-environment-27441156. Accessed August 2016.

22 See http://www.fernbankmuseum.org/explore/permanent-exhibitions/giants-of-the-mesozoic/. Accessed August 2016.

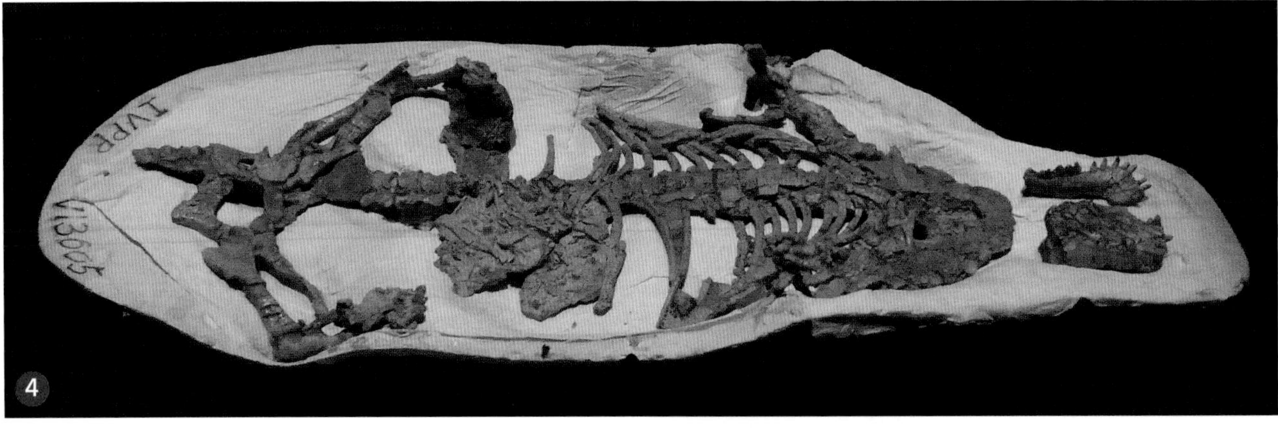

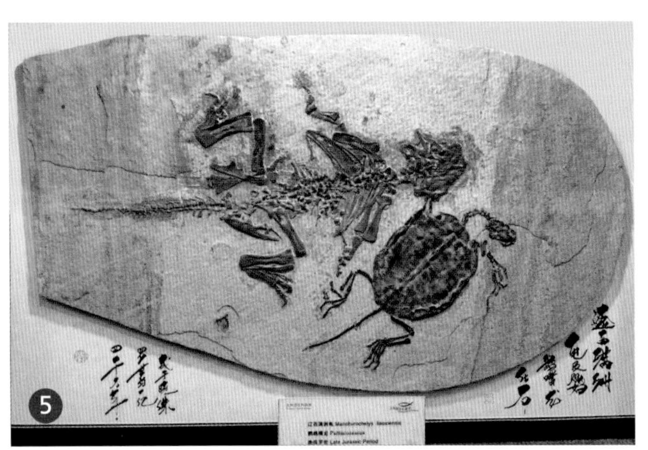

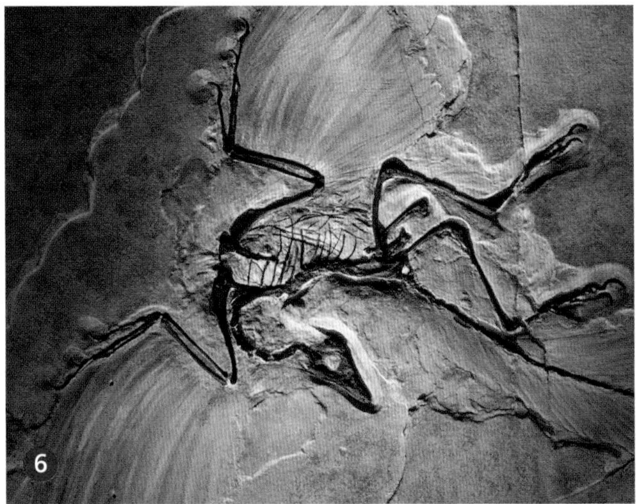

Dinosaurs buried alongside reptiles we have today

It is significant that the dinosaurs are buried in Cretaceous layers which also contain reptiles like crocodiles which we have today. There are a number of examples of burial sites where hundreds and sometimes thousands of dinosaurs are all buried together.[23] One example is in Spain where 8,000 fossils are buried, including several kinds of armor-clad plant-eating dinosaurs *(titanosaurs)* and, significantly, buried with them *at the same site* are turtles and crocodiles.[24] Dinosaurs are thus extinct reptiles which did not survive (figures 4, 5). They were buried in the Flood with many other creatures, and harsher conditions after the Flood have caused post-Flood descendants of dinosaurs to eventually die out.

A much overlooked fossil is that of the small dinosaur *Psittacosaurus* buried alongside a turtle *Manchurochelys liaoxiensis* in late Jurassic rock (figure 5). The turtle fossil is no different essentially than a modern turtle, and buried with it is a small dinosaur which we don't have today — it has become extinct. It was not changing to another

23 The discovery of thousands of Centrosaurus fossils near the town of Hilda, Alberta, is believed to be the largest bed of dinosaur bones ever discovered. The area is now known as the Hilda mega-bonebed. See M.J. Ryan, B.J. Chinnery-Allgeier, D.A. Eberth, P.J. Currie, and P.E. Ralrick, *New Perspectives on Horned Dinosaurs* (Indiana University Press, May 2010). See also http://www.edmontonsun.com/news/canada/2010/06/18/14439136.html. Accessed August 2016.

24 "Massive Dinosaur 'Graveyard' Discovered in Spain," *National Geographic* Dec. 10, 2007; http://news.nationalgeographic.com/news/2007/12/071210-dinosaur-grave.html .

4 Mammal *Repenomamus robustus* fossil with remains of *Psittacosaurus* in its stomach *(Photo courtesy of Kumiko, Tokyo, Japan. Wikimedia)*

5 Fossils of the small dinosaur *Psittacosaurus* (left) with the turtle *Manchurochelys liaoxiensis* (right) in late Jurassic rock at the Shanghai Ocean Aquarium *(Photo courtesy of Christopher, Tania, and Isabelle Luna. Wikimedia)*

6 *Archaeopteryx lithographica*, found in the Jurassic Solnhofen Limestone of southern Germany *(Wikimedia)*

creature! And yet conventional evolutionary ideas have mammals appearing many millions of years after dinosaurs. Another example is a mammal fossil that has been found with a baby *Psittacosaurus* in its stomach (figure 4). So, mammals were co-existent with dinosaurs.

Dinosaurs to birds?

Some evolutionists try to suggest that some dinosaurs are ancestral to birds based on the analysis of the group of dinosaurs with a bird-hipped pelvic structure. But as David Menton states, "... the bird-hipped dinosaurs, including such huge quadrupedal sauropods as *Brachiosaurus* and *Diplodocus*, are even less bird-like than the lizard-hipped, bipedal dinosaurs such as theropods"[25]

Evolutionists therefore seek other evidence to make the claim that birds have evolved from dinosaurs. They assert that the three-fingered hand is evidence of bird-like feet in dinosaurs. But the evidence is that the digits in theropod dinosaur feet are digits 1 (digit 1 is the thumb), 2, and 3, while birds utilize digits 2,3,and 4. So this evidence does not fit their hypothesis either.[27]

Furthermore, breathing in birds is very different from reptiles and involves a continuous flow lung which is not powered by a diaphragm which all reptiles and mammals have, but instead is powered by moving the breastbone (sternum) — see figures 1 and 2 on pages 56–57 where this system is described in greater detail. Unidirectional continuous flow lungs may yet be found in alligators, but the important difference is that alligators and indeed all reptiles (just as in mammals) have diaphragms and birds do not.

It is claimed that some ancestral dinosaurs were warm blooded and that these had feathers. However, claims of feathered dinosaurs are not convincing. The reader is referred to references 15 and 27 for an in-depth discussion of these

25 D. Menton, "Did Dinosaurs Turn into Birds?" chapter 2 of "Dinosaurs" Pocket Guide, Answers in Genesis, 2010, p. 41–53. See also A. Feduccia et al., "Do Feathered Dinosaurs Exist?" *Jnl. Morphology* 266, 125–166, 2005.

7 Dinosaur carving near Angkor Wat, in decorative walls of Ta Prohm temple, Siem Reap, Cambodia, built by King Jayavarman VII in the late 1100s. Most likely this is a carving of a stegosaurus. *(Wikimedia)*

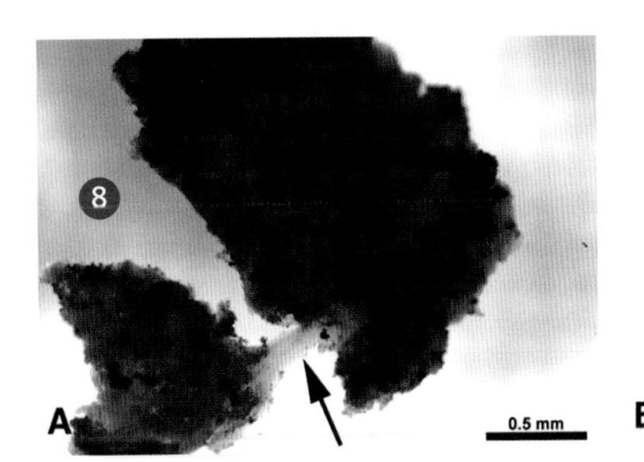

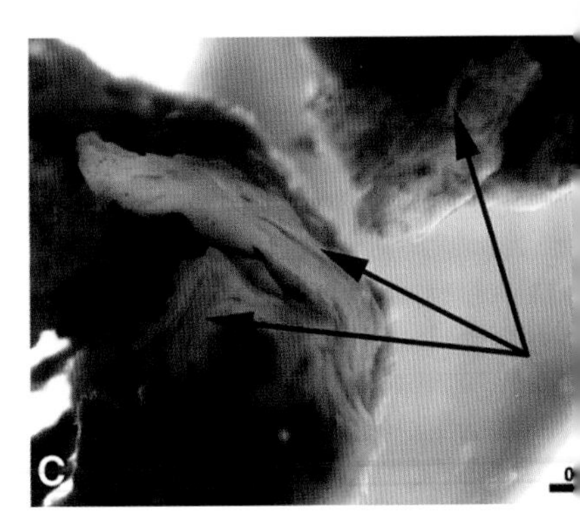

8 Soft unfossilized tissue found by Mary Schweitzer in *Tyrannosaurus Rex* fossilized femur. (Image from ref. 30. *Used courtesy of AAAS Science under copyright license 3944721100864*)

suggestions. Dinosaurs had scales (figure 2) and all the indications are that, as with all reptiles, they were cold-blooded creatures. Feathers would hinder the heating process necessary to warm a reptile's blood during the day. There are clear examples of genuine feathers fossilized in the rocks, but where these are found they are of extinct birds, not dinosaurs. The most famous example is *Archaeopteryx* (figure 6), which has fully developed flight feathers and was no halfway creature. That it had teeth and a bony tail with hooks on the wings, simply means that some birds in the past had teeth and bony tails — and the Hoatzin bird found in South America today has hooks on its wings.

One of the most intriguing discoveries has been that of drawings and carvings of large animals distinctly dinosaur like. There is a temple (thought to be approximately 900 years old) near Angkor Wat in Cambodia that has stone carvings on its walls of various creatures. Most people admit that one of these is a stegosaurus (figure 7). Such carvings and drawings[26] indicate that people knew of these creatures up to a few hundred years ago and there are historical accounts of men encountering large creatures.[27] Though most of these creatures were buried in the Flood, there would be two of each kind which survived and multiplied in the world after the Flood. It is likely that the post-Flood conditions led to their demise eventually because of the different nature of the new world, with the drying out of swamps such as the Sahara and inner Australia, which became deserts.

Soft tissues found inside fossilized dinosaur bones

To her amazement, while she was at Montana State University in 2005, Mary Schweitzer found soft tissue in the femur bone of a fossilized tyrannosaurus rex dinosaur (figure 8). Even the blood vessels were evident in this example of unfossilized material.[28] Many other similar finds have been made since this and, notably, Mark Armitage has found soft tissue in the horn of a triceratops fossil.[29] All these findings underline that these burials were not 65 million years ago, because such soft tissue could not survive that length of time even in frozen conditions underground.

Ichthyosaurs in the process of giving birth

We have seen many examples of fossils showing rapid burial — butterflies, dragonflies, fish in the process of eating another fish (see pages 79, 83, 91,

26 There are many other such drawings, including an inscription in brass of what appears to be two long-necked sauropods on a grave of Bishop Bell (1410–1496) in the nave of Carlisle cathedral; http://www.apologeticspress.org/APContent.aspx?category=9&article=3533. Accessed August 2016.

27 W. Cooper, *After the Flood* (New Wine Press, 1995). See chapter 10, "Dinosaurs from Anglo-Saxon and Other Records," and chapter 11, "Beowulf and the Creatures of Denmark." Both these chapters have lists of references to sightings of and attacks by large dragon-like reptiles both in Britain and Europe. Also, Bodie Hodge, "Dragon Legends — Truths Behind the Tales," *Answers Magazine*, Answers in Genesis, October 2011; https://answersingenesis.org/dinosaurs/dragon-legends/dragon-legends-truths-behind-the-tales.

28 M.H. Schweitzer, J.L. Wittmeyer, J.R. Horner, and J.K. Toporski, "Soft-Tissue Vessels and Cellular Preservation in Tyrannosaurus rex," *AAAS Science*, 307 (5717), 1952–1955, March 25, 2005.

29 M. Armitage, and K.L. Anderson, "Soft Sheets of Fibrillar Bone from a Fossil of the Supraorbital Horn of the Dinosaur Triceratops Horridus," *Acta Histochemica* 115, 603–608, 2013.

95, and in particular figure 2, page 186). Sometimes dinosaur burials are found by the thousands. Perhaps one of the most remarkable examples is of a creature in the process of giving birth. Seagoing creatures that are now extinct are also found in the fossil record. Ichthyosaurs were evidently something like our dolphins and porpoises today in that they gave birth to live young at sea as is evidenced by the very striking fossil shown in figure 9 of an ichthyosaur actually giving birth to one embryo and having two other embryos visible as well.[30] A snapshot in the middle of birth is strong evidence indeed of rapid burial. These and other examples all provide evidence for a global Flood. The message of the rocks is of a changed world.

30 R. Motani, Jiang D-y, A. Tintori, O. Rieppel, Chen G-b, "Terrestrial Origin of Viviparity in Mesozoic Marine Reptiles Indicated by Early Triassic Embryonic Fossils," PLoS ONE 9(2), 2014.

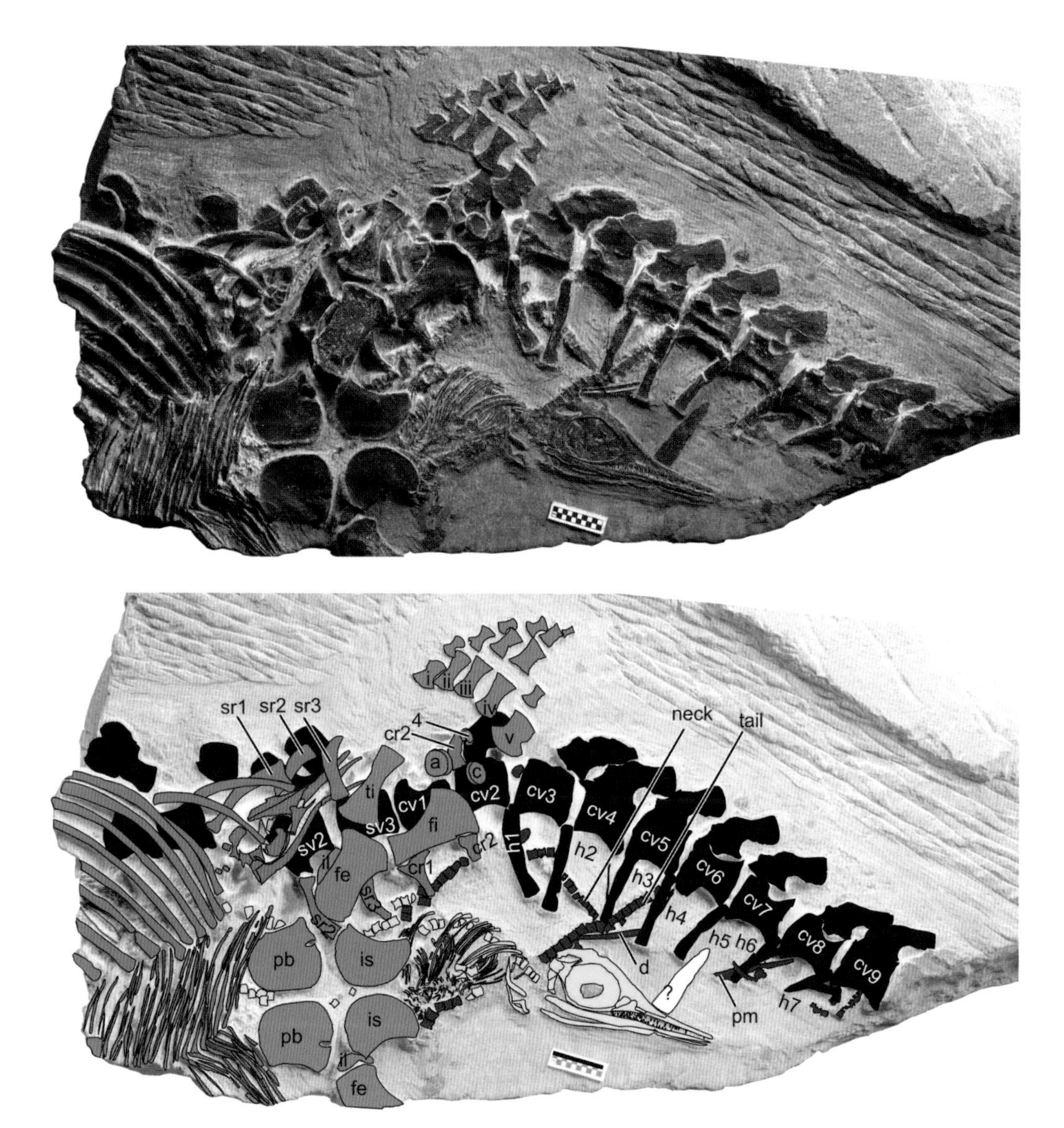

9 Fossil of ichthyosaur actually giving birth — Chaohu, Anhui, China — from reference 32. Embryos 1 and 2 are in orange and yellow, respectively, whereas bones of one just born are in red.

Radioactive Dating and What of Carbon 14?

Various methods of dating are claimed to support a very old Earth, but are the assumptions behind these methods reliable?

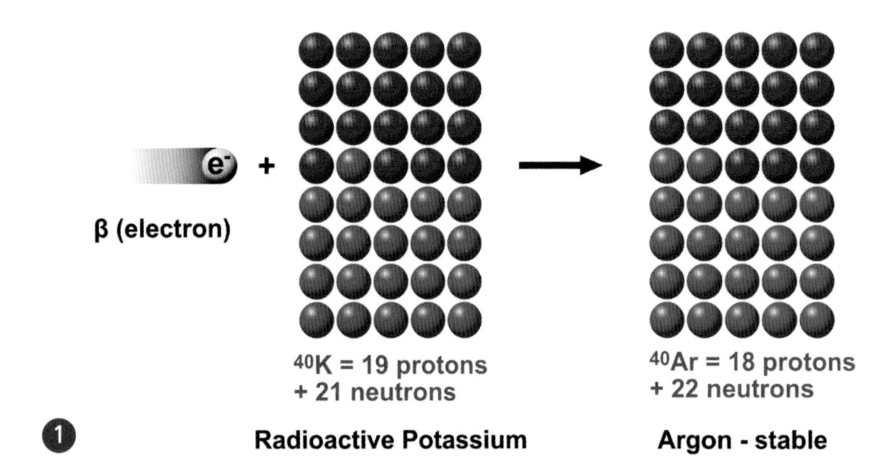

1 Potassium Argon dating: one unstable atom of 19 protons and 21 neutrons undergoes a change where one proton absorbs an electron to become a stable atom of Argon with 18 protons and 22 neutrons — 11% of radioactive Potassium atoms decay into Argon by this mechanism involving electron capture.

Most scholars today think that the rocks have been proved to be millions of years old, but when all the evidence is examined carefully, this is not proven at all. The media have so pushed this that the average person is left bemused concerning the age of rocks, because it does involve matters like radioactive dating which most find too hard to understand. When one looks at the method there are huge assumptions which are critical to all the conclusions.

How does radioactive dating work?

In some of the volcanic rocks there are minerals that are radioactive. These gradually decay to produce a different element. So, for example, (figure 1) Potassium 40 decays to Argon 40 because the Potassium (as all chemical elements) is defined by the number of protons in the nucleus, which for Potassium is 19, along with (normally) 20 neutrons. The protons and neutrons have almost exactly the same weight, so natural Potassium normally has an atomic weight of 39 (using the proton as one unit). The radioactive version of Potassium has one extra neutron so its atomic weight is 40. It is still Potassium, because it starts off with 19 protons, but this version of Potassium with 21 neutrons is unstable, and so one of the protons absorbs an electron which changes the proton into a neutron causing it now to become a different element with 18 protons and 22 neutrons. This element is the stable form of the gas Argon.

This "radioactive decay" takes place very slowly and its rate can be measured in the laboratory. So by measuring the amount of radioactive Potassium (the parent element ^{40}K) in rocks today and the trapped Argon (the daughter element ^{40}Ar), and knowing the decay rate, one can then infer the age of the rock. However, there are very important assumptions involved in this.

The analogy of a sand timer (figure 2) can be helpful in understanding the assumptions in radioactive decay calculations. First, one cannot be sure as to what the initial values of parent and

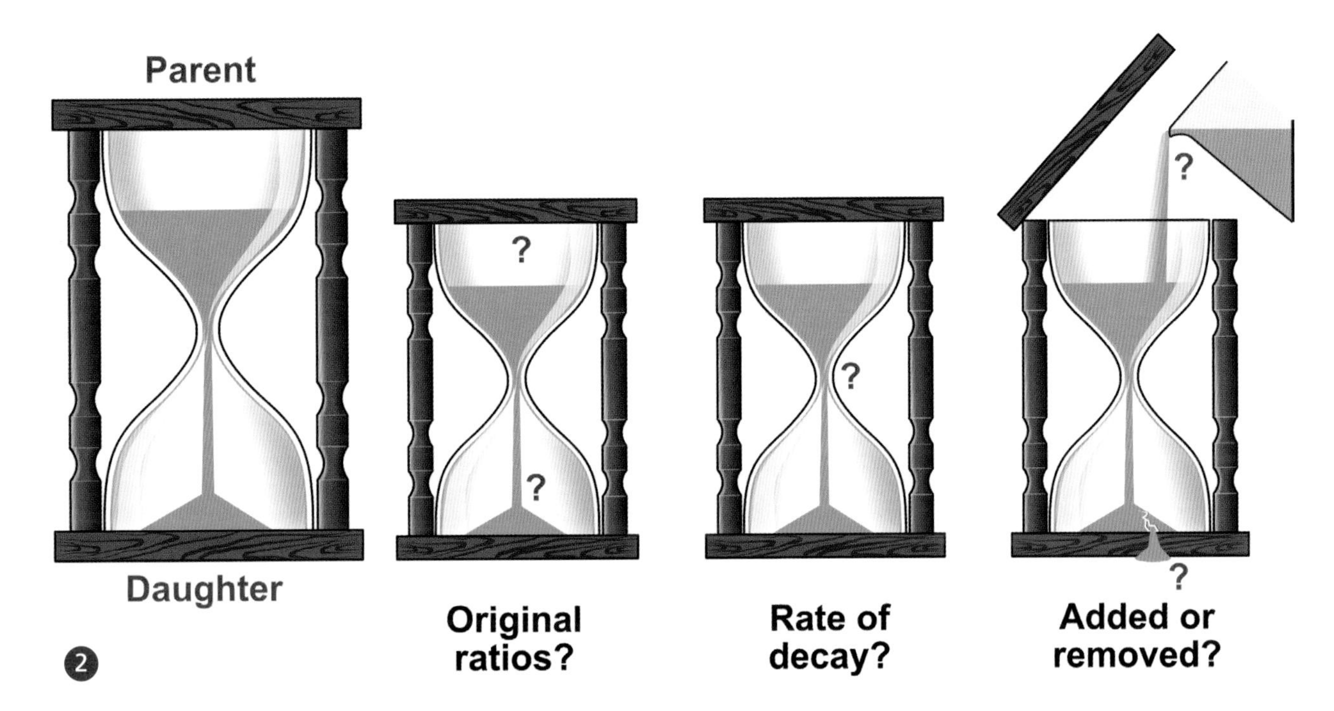

2 Understanding the principle of radioactive dating using the sand timer analogy

daughter element were. In our illustration (figure 2) using the sand-timer, we don't know how much sand was in the top in the first place, nor do we know whether some of the sand in the bottom might have leaked out through a crack in the glass. So in the Potassium-Argon example, one does not know how much ^{40}K (radioactive Potassium) and trapped ^{40}Ar (Argon) were present at time zero (when the rock was formed). There are clear cases of ^{40}Ar being in the original rock as the magma cooled.[31] Furthermore, there is some evidence that the rate of decay might possibly be affected by changes in temperature, though this is disputed.[32] Also, since the volcanic rock is porous, there is the possibility that some of the parent element could have been leached out. This would then have a major effect on the calculated age, making it much less than assumed.

Concerning the decay rate and whether it is constant, surprising evidence has emerged indicating that even the sun affects radioactive decay rates[33] and no one yet knows what causes these seasonal variations.

A team of research scientists has considered the possibility of accelerated radioactive decay operating in the past, particularly at the time of the

31 A. Snelling, "Radioactive Dating: Research Confirming the Biblical Record," Genesis Agendum Occasional Paper 8, 2009. See page 5 and discussion of anomalous radioisotopic dating results.

32 B. Limata et al., "First Hints on a Change of the 22Na β+-decay half-life in the metal Pd," *Eur. Phys. J.* A 28, 251–252, 2006; also see B. Wang et al., "Change of the 7Be Electron Capture Half-life in Metallic Environments," *Eur. Phys. J.* A 28, 375–377, 2006, and J.R. Goodwin, V.V. Golovko, V.E. Iacob, and J.C. Hardy, "Half-life of the Electron-capture Decay of 97Ru: Precision Measurement Shows No Temperature Dependence," Physical Review C 80, 045501, 1–6, 2009. Dr. Tas Walker discusses the earlier findings in T. Walker, "Radioactive Decay Rate Depends on Chemical Environment," *Journal of Creation* 14(1):4–5, April 2000.

33 Jenkins et al., "Additional Experimental Evidence for a Solar Influence on Nuclear Decay Rates," *Astroparticle Physics* 37, 81–88, 2012; see also B. Thomas, "The Sun Alters Radioactive Decay Rates," ICR Creation Science Update, Sept 3, 2010; http://www.icr.org/article/sun-alters-radioactive-decay-rates.

3 An ammonite from cretaceous layers, Northern California, is shown encased in fossilized wood. *(Photo courtesy of Al Franklin, Redding, California)*

Flood.[34] For a good summary of the issues the reader is referred to the footnotes.[35]

34 L. Vardiman, A.A. Snelling, and E.F. Chaffin, "Radioisotopes and the Age of the Earth: a Young-earth Creationist Research Initiative, Institute for Creation Research and Creation Research Society," 2000. Dr Russ Humphreys suggests a possible mechanism for changing radioactive decay rates — see D.R. Humphreys, "Accelerated Nuclear Decay: a Viable Hypothesis?" see p. 333– 379.

35 A. Snelling, "Radioactive Dating : Research Confirming the Biblical Record," Genesis Agendum Occasional Paper 8, 2009. On p. 15–16, see discussion of possible mechanisms concerning how decay rates could alter.

G. Barnard, A. McIntosh, and S. Taylor, "Origins — Examining the evidence," Truth in Science, 2011; see section on radiometric dating by Taylor on p. 65–68.

Large discrepancies in dating schemes

In the light of the questionable assumptions in radioactive dating, we are not surprised to learn that on a number of occasions large anomalies are found. Rocks gathered from an eruption of Mt. Ngauruhoe in New Zealand gave a Potassium-Argon date of

V. Cupps, *ICR Acts & Facts*, Radioactive Dating: Part 1, 43(10), 2014; Part 2, 43(11), 2014; Part 3, 43(12), 2014; Part 4, 44(2), 2015; Part 5, 44(3), 2015; Part 6, 44(4), 2015; Part 7, 44(5), 2015; Part 8, 44(6), 2015. See http://www.icr.org/article/8348 and follow links for all eight articles.

A.A. Snelling, *Earth's Catastrophic Past — Geology, Creation and the Flood*, Volumes I and II (Institute for Creation Research, 2009).

270,000 to 3.5 million years, when the rocks are known to be less than 60 years old![36]

Another example (figure 3) is of an ammonite from cretaceous layers that is encased in fossilized wood. Snelling[37] found significant Carbon 14 in the fossilized wood, which is obviously wrapped round the ammonite and which dates the find to thousands of years old, whereas, according to evolutionary timing, the cretaceous layers date to 112–120 million years old.

Examples of granites have been found where radioactive Uranium decaying to Lead produces Helium trapped inside the crystals within the granites. When the Helium level is measured, the level is consistent with thousands of years even though the Uranium-Lead dating scheme itself implies the granites are 1.5 billion years old.[38] The discrepancy in each case is colossal and these anomalies confirm that the assumptions of the radioisotope schemes are in error.

Carbon 14 dating

Carbon 14 dating (figure 4) deals with the radioactive decay of the isotope ^{14}C of Carbon (8 neutrons and 6 protons) to Nitrogen (^{14}N – 7 neutrons and 7 protons), which has a much faster decay rate than for ^{40}K decaying to ^{40}Ar or Uranium decaying to Lead. The half-life is the time taken for a radioactive substance to reach half of its original value. For ^{40}K – ^{40}Ar this is very long — 1,251 million years. But for ^{14}C – ^{14}N it is only 5,730 years. This means that in 5,730 years, any ^{14}C present will be at half its original value. In 10 half lives it will be at approximately 1/1000th of its original value. By 20 half lives (just under 115,000 years) approximately a millionth of the original ^{14}C would be present, which means that one can regard all the ^{14}C to have vanished.

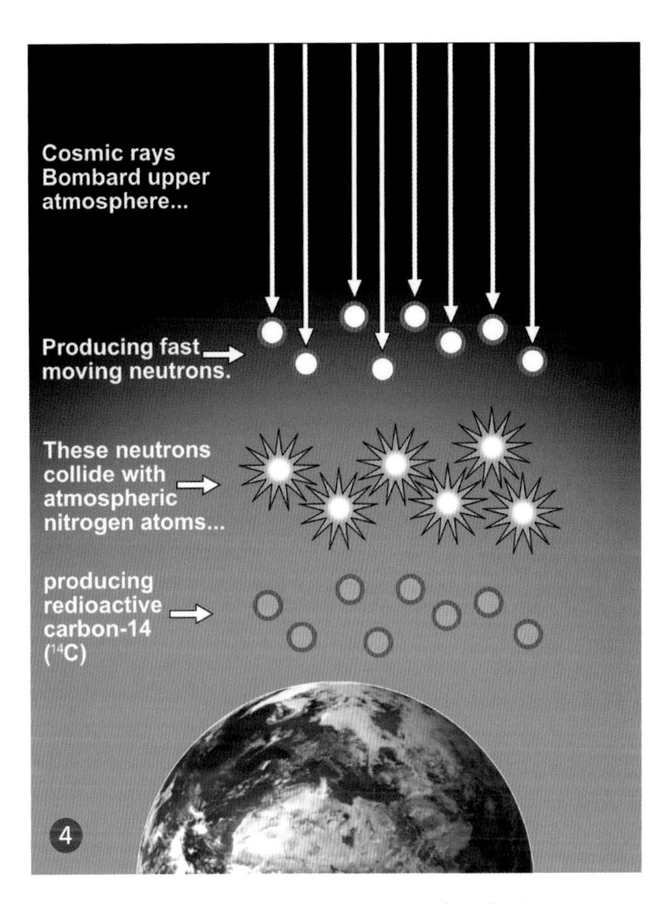

4 Carbon 14 produced in the upper atmosphere by Cosmic radiation

The remarkable finding is that ^{14}C *has* been discovered at significant levels in coal, fossilized wood, and even diamonds.[39] This means that whatever the initial value of ^{14}C, these layers must be much less than the upper limit of approximately 100,000 years, which is much less than the millions of years usually assigned to them by other schemes.[40] In the same way as with Uranium-Lead and Potassium-Argon dating schemes, one does

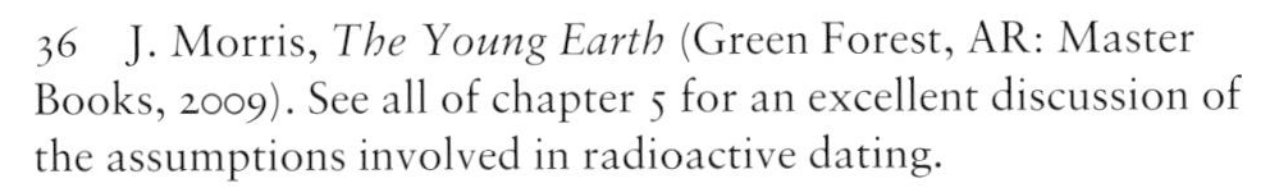

36 J. Morris, *The Young Earth* (Green Forest, AR: Master Books, 2009). See all of chapter 5 for an excellent discussion of the assumptions involved in radioactive dating.

37 A.A. Snelling, "Radiocarbon Ages for Fossil Ammonites and Wood in Cretaceous Strata near Redding, California," *Answers Magazine*, AiG, December 2008; https://answersingenesis.org/geology/carbon-14/radiocarbon-ages-fossils-cretaceous-strata-redding-california/.

38 D.R. Humphreys, "Helium Retention in Zircon Crystal," chapter 4 of D.DeYoung, ed., *Thousand not billions* (Green Forest, AR: Master Books, 2010).

39 J. Baumgardner, "Carbon-14 Dating," chapter 3 of D. DeYoung, ed., *Thousand Not Billions* (Green Forest, AR: Master Books, 2010).

40 A. Snelling, "Radioactive Dating : Research Confirming the Biblical Record," Genesis Agendum Occasional Paper 8, 2009. See p. 15–17 for discussion of Carbon 14 dating schemes.

not know the original Carbon 14 levels, but the maximum age possible is nowhere near millions of years. Carbon 14 traces in fossilized wood, coal, and diamonds are a big indicator that the Earth is certainly not millions of years old.

Carbon 14 (^{14}C) is produced (figure 4) by cosmic radiation causing a neutron to collide with a Nitrogen molecule (^{14}N) in the upper atmosphere and replacing a proton which is ejected, so that the isotope ^{14}C nucleus is produced. It is likely that the pool of ^{14}C in the upper atmosphere was lower in the past due to less cosmic radiation reaching the Earth. This would lower the starting condition of ^{14}C in the atmosphere and consequently the ages calculated would be measured in biblical timescales of thousands not tens of thousands of years.

Further indications of a young Earth

Apart from radioactive and Carbon 14 dating, there is the strong evidence of the layers themselves, such as in the Grand Canyon (figure 5), which indicate rapid catastrophic processes have been involved.

Furthermore, the very presence of large coal seams across the world (figure 6) are also a strong

5 The Grand Canyon in Arizona from the South Rim at sunset — the horizontal layers are strong evidence of the Flood with subsequent near vertical carving out of the Canyon by fast-flowing water after the sediments had been laid down.

6 Coastal exposure of the Sydney Mines Point Aconi Seam (bituminous coal; Pennsylvanian), Nova Scotia, Canada
(Photo by Michael C. Rygel, Wikimedia)

indicator that there has been a global cataclysm. The production of coal with high carbon content (called "high rank" coal) is not possible by simply leaving vegetation to gradually decompose. The most one can obtain is peat. The reason for this is that when open to the air, the vegetation not only disintegrates into its component elements, but also the carbon combines with the oxygen in the air to produce CO_2. There is very slow combustion.

However, as soon as you exclude the air, then the disintegration (called pyrolysis) to Carbon, Nitrogen, and other elements takes place without the production of CO_2. This is what we find in major high rank coal deposits across the world on every continent (including Antarctica), and is strong evidence of catastrophic processes. In some locations such as the Latrobe Valley Coal Measures, Victoria, Australia, the seam is 2,300 feet (700 m) thick, 190 miles (300 km) long, and 190 miles (300 km) wide. Offshore, the thickness increases to 3 miles (5 km). Though Ager[41] was no friend to the creation/Flood position, he acknowledged that the evidence of catastrophe in the Earth's sedimentary layers, and in particular the coal fields, was undeniable.

Further evidence of a young Earth and a catastrophic flood was given in the previous section with the soft tissue found inside fossilized dinosaur bones and the ichthyosaurs that died in the process of giving birth (pages 194–195).

41 D. Ager, *The New Catatstrophism — the Importance of the Rare Event in Geological History* (Cambridge University Press, 1995).

13 The Six Days of Genesis

The Genesis account of creation is intended to be taken as history. This is consistent with the style of writing, the discoveries of observational science, and a correct understanding of the biblical text.

Genesis and the Style of Writing

The genre of Genesis chapter one — the style of writing intended by the original writer and how he would expect it to be understood by the readers — is history and not poetry. This has been understood by the great majority of Hebrew scholars and Bible commentators from the earliest times.[1]

Many of the early church fathers believed that the six days were literal but also symbolic of 6,000 years of Earth's history to the end of time. They therefore almost all believed in a young Earth. Irenaeus of Lyons is an example of many: "For in as many days as this world was made, in so many thousand years shall it be concluded."[2] Augustine of Hippo, while accepting the total time span of Earth to be 6,000 years, believed that creation was instantaneous.[3]

However, Martin Luther and John Calvin are representative of the Protestant Reformers in the 16th century and these were men of significant scholarship in Hebrew. Luther wrote, "We assert that Moses spoke in the literal sense, not allegorically or figuratively."[4] Calvin even suggests

1 David Hall overviews the history of exegesis of Genesis 1–3 in *Coming to Grips with Genesis* (Green Forest. AR: Master Books, 2008), p 53–78. See also Stephen Boyd on p 176. Unfortunately, Hugh Ross in *Creation and Time* (Colorado Springs, CO: NavPress, 1994) in defending an evolutionary interpretation of Genesis, misunderstands the early church fathers, particularly Augustine, and chooses to ignore entirely the united evidence of the Reformers, Puritans and 18th-century biblical scholars!

2 Heresies 5.28:3.

3 *City of God* 12 (10).

4 *Luther on the Creation,* "A Critical and Devotional Commentary on Genesis," edited by Prof. John Nicholas Lenker, based on the translation by Dr Henry Coles from the original Latin (Minneapolis, MN: Lutherans in all Lands Co., 1904), p 40–41.

that if we question God's creation in six days "We slightingly pass over the infinite glory of God, which here shines forth."[5] The Westminster Confession of Faith, which was the accepted theology of the Puritans throughout the 17th century and beyond, states that God created all things "in the space of six days."[6] The great Bible commentators of this period — Matthew Henry, John Gill, and Thomas Scott — all believed in a young Earth and six-day creation.

Johann Keil, an outstanding 19th-century scholar of Hebrew, in his commentary on these early chapters, concluded, "They bear the marks, both in form and substance, of a historical document in which it is intended that we should accept as actual truth, not only the assertion that God created the heavens and the Earth, and all that lives and moves in the world, but also the description of the creation itself in all its several stages."[7]

Professor James Barr, Regius Professor of Hebrew at Oxford from 1978 to 1989, claimed, "So far as I know, there is no professor of Hebrew or Old Testament at any world-class university who does not believe that the writer(s) of Genesis 1–11 intended to convey to the readers the idea that creation took place in a series of six days which were the same as the days of 24 hours we now experience."[8]

Until the arrival of the theory of evolution, almost all Hebrew and biblical scholars understood the genre of the first chapters of the Bible as a record of historical fact. Even today, most evolutionists who try to be loyal to the Bible admit that Adam and Eve were real figures of history. Their dilemma is, where does the genre of Genesis suddenly change?

An important hermeneutic principle is that we must not superimpose what we want a passage to mean (from supposed evolutionary thinking), before understanding what the text states, what the original writer intended, and how the original readers understood it.

Genesis and Observational Science

Throughout this book we have considered the scientific facts as seen today — this is "observational science." It is more reliable than an evolutionary "theory" — what we think may have happened. The evidence of observational science is entirely consistent with creation followed by a worldwide Flood.

The evidence of irreducible complexity is equally consistent with all things created complete and "ready to go." Things cannot function unless they are complete. A part-evolved brain, eye, or ear would be useless. A part-evolved wing of a dragonfly, bee, or butterfly is of no value. Pluck two leading wing feathers from a bird and it is grounded. Unless salmon were complete with the genetic code for migration from the very beginning, they would never have survived. And so the list continues. The full DNA information must be coded in right from the start.

Similarly, we have seen that all observable scientific evidence of the stars and planets indicate a solar system that cannot be billions of years old.

Unfortunately, the mantra of evolution has been upgraded from a hypothesis to a theory without any certain scientific proof. It has become a widespread faith system based on historical science, because the only alternative is creation — and in a very short space of time — by a divine Being of infinite power and wisdom.

This is precisely the claim of the creation account in Genesis. At the close of each day, God declared that all he had made was "good." At the end of the creation week it was "very good" — in the Hebrew that is an absolute superlative. At this point there was no disease, cruelty, or death on planet Earth.

We should not forget God's challenge to Job: "Where were you when I laid the earth's foundation?" (Job 38:4; NIV).

5 *Calvin's Commentaries* (Grand Rapids, MI: Baker Book House, 1979, original 1563), Vol. 1, p. 78.

6 *The Westminster Confession of Faith* (1646). 4:1.

7 *Keil and Delitzsch Commentary on the Old Testament* (1861) (Grand Rapids, MI: William Eerdman Publishing Company, 1976), Vol. 1, p. 37–39.

8 Personal correspondence with David CC Watson, April 23, 1984.

Genesis and the Meaning of the Text

For those who attempt a marriage between the Bible and evolution, there are two ways to handle the account in Genesis chapters 1 and 2. The first suggestion, that the style of writing is fiction or poetry and not history, has already been answered.

The second suggestion is that the "days" are not to be taken as literal 24-hour periods. However, there is no agreement on how they should then be interpreted. Some consider them to be the days on which God revealed His creation to Moses, to others they represent long periods of time to allow for evolution, and to others the whole account is cast in the framework of a poetic narrative. Some place gaps at various points in the text to allow for millions or billions of years. This complete disagreement of interpretations is the inevitable result of confusing that which is perfectly clear.

The primary use of the word *yom* (day) in Hebrew is a 24-hour period, so there has to be compelling reasons from within the text to understand it as a longer period of time.

When the Hebrew word for "day" (*yom*) is used with a number, and here you have "the first day," "the second day," and so on, it refers to a literal 24-hour period. Outside Genesis in the Old Testament there are almost 360 occasions when *yom* is used with a number, and with one single possible exception (Hosea 6:2) it always refers to a 24-hour period. One out of 360 can hardly challenge the primary use.

The reference to "evening and morning" cannot seriously be taken to refer to anything other than a 24-hour period. This phrase, used 19 times outside Genesis 1, never refers to a long period of time.

Genesis 1:14 refers to "day," "night," "seasons," "days," and "years"; it would be remarkable to select only the word "day" as non-literal in that list. The same word "day" occurs 17 times in the record of the global Flood (Genesis 7, 8), and no one suggests that in those instances it means anything other than the cycle of night and day.

In Exodus 20:11, the six creating days became the pattern for a six-day working week ". . . in six days the Lord made the heavens and the earth, the sea, and all that is in them, but he rested on the seventh day. Therefore, the Lord blessed the Sabbath day and made it holy." There can be no doubt that Moses and the nation of Israel understood creation taking place within one week.

Prof. John Skinner who, in the 19th century was a professor of Old Testament Language and Literature in Cambridge, insisted that the attempt to harmonize science and Genesis by claiming that "day" here means an aeon is "exegetically indefensible" and "opposed to the plain sense of the passage and has no warrant in Hebrew usage."[9] Skinner had no bias in favor of biblical inerrancy.

Similarly, Prof. S.R. Driver, one of the foremost Old Testament scholars of the early 20th century and Regius Professor of Hebrew at Oxford concluded, "There is no occasion to understand the word [day] in any but its ordinary sense."[10]

Prof. Leupold, a prominent Hebrew scholar in the middle of the last century, concluded: "There ought to be no need of refuting the idea that *yom* means period. Reputable dictionaries . . . know nothing of this notion."[11]

Those who believe in the authority of the Bible and who therefore accept the miracles recorded (including the account of Jonah, the virgin conception of Christ, and His literal Resurrection from the dead) and yet defend an evolutionary account of creation, have adopted historical science as their authority. One such writer admits this and adds "they may be wrong, for all I know."[12] That is a dangerous foundation for understanding the Bible.

There is only one reason to interpret the word "day" as a period of time, and that is the acceptance of the evolutionary hypotheses that the universe is billions of years old. There can be no other biblical, historical, or scientific reason.

God Himself has declared:

It is I who made the earth
and created mankind on it.
My own hands stretched out the heavens;
I marshaled their starry hosts.
(Isaiah 45:12; NIV)

9 S.R. Driver, A. Plummer, and C.A. Briggs, *The International Critical Commentary* (Edinburgh: T&T Clark, 1895), p. 5, 21.

10 S.R. Driver *The Book of Genesis* (London: Methuen and Co., 1905), p 26, 6.

11 H.C. Leupold *An Exposition of Genesis* (The Wartburg Press 1942. EP edition), p. 57.

12 C. John Collins. *Science and Faith: Friends or Foes?* (Wheaton, IL: Crossway Books. 2003), p. 250.

14 What Happened?

A beautiful universe, perfectly fitted for planet Earth, and planet Earth perfectly designed for all living creatures, is marred by death, destruction, and decay.

We have followed an awe-inspiring journey to discover the incredible diversity, complexity, and beauty of creation. Science and a rational mind must surely agree that planet Earth could not have evolved over billions of years by a vast series of cosmic and genetic accidents. The "irreducible complexity" that we have observed time and again demonstrates that all living creatures on the land, sea, and air could only exist and function when they were fully and elaborately formed.

But where does this lead us? First, the more that true science discovers about the awesome complexity of planet Earth and the universe, the clearer it points to a Creator of immense power and wisdom. Second, the sheer splendor and wonder of the natural world around us reveals that Creator as a God of kindness, who Himself loves beauty and design and wants us to enjoy a good relationship with Him on a beautiful planet. The evidence of His existence is so plain that unbelief has no excuse. As Paul wrote long ago "Since the creation of the world God's invisible qualities — his eternal power and divine nature — have been clearly seen, being understood from what has been made" (Romans 1:20).

"Very good"

The Bible comments that at the end of God's six days of creation all was "very good" (Genesis 1:31). This was not simply the best that God could do, but the very best that could be done — because God did it. There was no violence, pain, suffering, or death. It was a perfect creation, ready for the

> The creation record in Genesis 2 and 3 describes how the first man and woman enjoyed a profound and pure relationship with their Creator and all the benefits of a perfect creation.

residence of the climax of all God's work — a man and a woman made in the image and likeness of God Himself (Genesis 1:26). This "image and likeness" is the uniqueness of the human race in our ability to appreciate beauty and to discover, design, and invent in a way unknown to the animal world; to experience strong and deep emotions of love, friendship, and compassion; to be moral, understanding the difference between right and wrong, and to enjoy a meaningful relationship with the Creator Himself. All this is a reflection of the character of God.

The creation record in Genesis 2 and 3 describes how the first man and woman enjoyed a profound and pure relationship with their Creator and all the benefits of a perfect creation.

The whole universe was created for planet Earth. Stars and planets were made so that human beings have for millennia fixed times and seasons, and found their way across vast oceans and landscapes. The Earth itself was stored with rich minerals for humanity to use and enjoy, and with vegetation in a plentiful supply of fruit and vegetables with all their valuable nutrients for life. Even the animals were created in such a way as to be of maximum usefulness for mankind. In addition to all this, the universe was formed with a spectacular splendor of color, beauty, and order, and the Earth with a variety of living things that we have even yet to discover fully. After millennia of human existence, we are still finding creatures on the land, flowers in the field, fish in the sea, and stars and planets in the heavens that we never before knew existed.

Without question, the whole universe was very good and just right for the first man and woman — Adam and Eve.

Something Went Terribly Wrong!

Crashing into this magnificently beautiful creation, one tragic event ruined everything.

The Christian explanation for what went wrong is found in the account in Genesis which describes how the Creator gave just one instruction for perpetual happiness: obedience to His wise and benevolent rule. When Adam and Eve were deceived into believing that God did not really mean what He said, the instruction was disobeyed and as a result — disaster (Genesis 2:17, 3:1).

This account of the "Fall" is not a legend. It is reality and the explanation for all that follows in the history of the human race. Certainly Jesus and His disciples recognized it as an historical account of the origin of human life, sin, and suffering (Matthew 19:4–5; 1 Timothy 2:13–14; Hebrews 1:10). There is no good reason to doubt it.

As a result of the Fall, creatures became "red in tooth and claw," the climate and terrain across much of planet Earth became unforgiving and inhospitable and even the vegetation fought against man's efforts to control it, disease and death entered, and the human race turned against its God in rebellion and against itself with greed, lust, and cruelty. Evil increased so much that God brought a worldwide Flood, while saving one family of eight people and pairs of all the land animals in an Ark. He restarted the world from this family but evil gradually grew again in the human race. Now greed and cruelty have extended across our world today. God's standard set out in the Bible teaches the human race how to keep relationships pure and peaceful, how to care for the environment, and how to live in friendship with God. Sadly, by ignoring God, death and disease, pain and suffering, violence and cruelty have become an overriding part of the human story.

The Apostle Paul expressed it bluntly when he stated that "Sin entered the world through one man, and death through sin, and in this way death came to all men, because all sinned." More than this, the whole creation is in "bondage to decay" (Romans 5:12, 8:21). That last claim is undeniable.

From paradise to pain

When God declared to Adam "Cursed is the ground because of you" (Genesis 3:17), He introduced some radical change as a result of Adam's disobedience, and implied that the Curse would affect the whole of creation. The warning in the following verse that "It will produce thorns and thistles" shows that this was not part of the six days of creation. So, how did it all change?

The most likely explanation is that God withdrew some of His protective care over His perfect creation, and this allowed an inevitable deterioration. All things, if left to themselves, will become degraded; everything we know, without any exception, is wearing out. This is exactly how the Bible sees it in Hebrews 1:10–12.

God did not suddenly inflict death and disease on Adam and Eve, but withdrew His protective care. This allowed mutations within living creatures and harmful bacteria to invade the whole of creation. Mutations have never been observed to increase information or create a new structure. The vast majority cause loss (not gain) of function. Though many bacteria are necessary and useful, they are deleterious when out of control. Similarly, the peaceful order among the animal kingdom descended into a battle for supremacy and jaws and teeth designed for cracking fruit and nuts now tore into flesh.

The whole world and universe is spoiled and is waiting for the Creator to set everything right. This is precisely what Paul means in Romans 8:21–22 where he reminds us that when Christ comes again "the creation itself will be liberated from its bondage to decay and brought into the glorious freedom of the children of God."

In a broken world, we can still see the imprint of God's order, design, and beauty. But the order and design are broken, beauty is tarnished, harmony is shattered, joy is compromised, and God's bountiful provision is manipulated by the human race to its own destruction.

Suffering in a broken world

To Job, a man whose story is recorded in the Bible and who questioned the fairness of his own extreme personal suffering, God responded, "Would you discredit my justice? Would you condemn me to justify yourself?" (Job 40:8; NIV).

But a few questions stand out:

WHY DIDN'T GOD stop the human race from sinning in the first place?

God gave Adam the freedom to choose in order to treat him with the dignity of being created in God's own image. Tragically, he squandered that freedom.

WHY DIDN'T GOD simply forgive Adam and overlook that first disobedience?

Then the human race would never learn anything. We would assume that we could break the natural and spiritual laws of God without consequences.

WHY DO THOSE CONSEQUENCES have to be so severe and far reaching?

Because every breach of natural and moral laws has outcomes that affect others. Pain, like alarm clocks that steadily rise in their decibels, warns us that something is wrong and needs attention. We raise our voice to the disobedient child, and we raise the penalty for repeated crimes.

> Suffering distinguishes between the human race and the kingdom of animals. The courageous triumph over personal disability and the human response to an appalling tragedy, each impressively illustrates the uniqueness of humanity

Is there value in suffering?

SUFFERING teaches us the value of obedience to God and the catastrophe of disobedience. If the whole world followed the teaching and example of Jesus Christ, the greater proportion of misery and pain would be avoided. But if we choose not to follow Him, we reap the consequence.

SUFFERING teaches us to be more careful. If God intervened to prevent every accident, pilot error, industrial negligence, like lazy children with a doting parent we would never learn. God's rules in the Bible teach us to care for our neighbor and our environment. When we ignore those rules we learn the hard way.

SUFFERING distinguishes between the human race and the kingdom of animals. The courageous triumph over personal disability and the human response to an appalling tragedy, each impressively illustrates the uniqueness of humanity. Humans spend days searching for survivors, but there is nothing like this among the animals. The world is full of suffering, but it is also full of courage and compassion to overcome it.

SUFFERING demonstrates the value of Christian faith. While many claim that they cannot believe in God in the face of suffering, they are challenged by the millions among every nation and language who suffer intensely and still trust in God.

SUFFERING is a warning to the human race. In our arrogance we often behave as if we are "master of our fate and captain of our soul." A devastating tornado, volcano, tsunami, or earthquake compels us to acknowledge how frail the human race is and how dependent upon our Creator we are. The literary scholar C.S. Lewis wrote, "God whispers to us in our pleasure, speaks in our consciences, but shouts in our pains: it is his megaphone to rouse a deaf world."[1]

SUFFERING, like aging, is intended to make us think about the fragility of life and accountability to God. If we choose not to think, then we have only ourselves to blame.

What Christ suffered shows that God was not unfeeling to the sufferings of the human race made in His image. He was prepared to endure deep suffering which was not only a physical, but a profound spiritual agony in our place (Isaiah 53:5–6)

Christians do not have all the answers — there are many secret things that belong to God (Deuteronomy 29:29). However, Christians do have reasonable answers to the problem of suffering.

For those who deny the Creator, there is no ultimate purpose either in the world around us or in suffering. They simply exist, and we each make of them whatever we will.

The ultimate purpose of the life, death, and Resurrection of Jesus Christ is to restore men and women to friendship with God so that they might enjoy their Creator forever.

1 C.S. Lewis. *A Grief Observed*. First published in 1961 by Faber and Faber under the pseudonym N.W. Clerk.

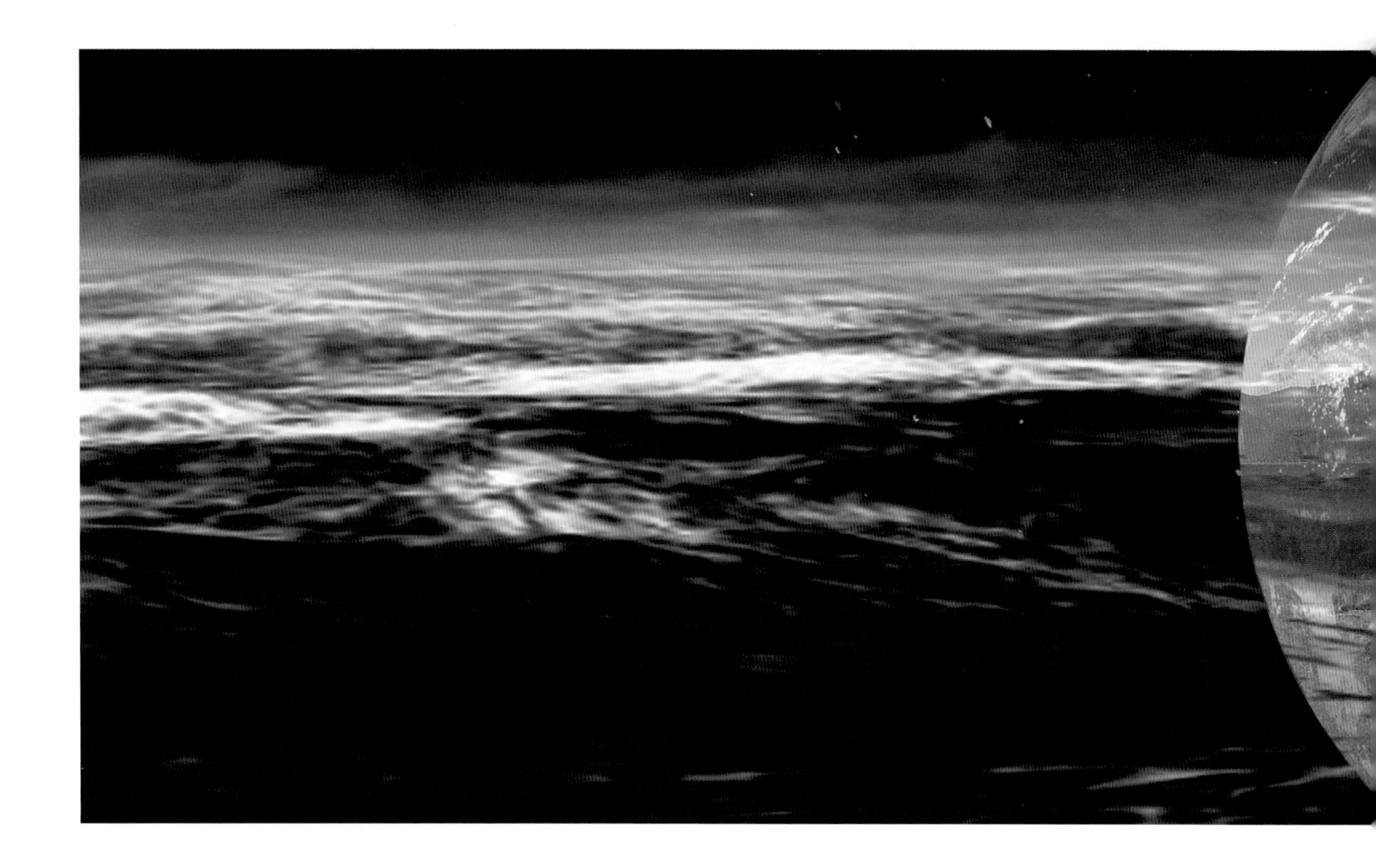

The meaning of life

At the start of this book we said that it would not close until we had answered the question "What is the purpose of it all?" Very simply, the powerful, wise, and loving Creator wants us to enjoy a good relationship with Him and enjoy His beautiful creation both now and for ever.

It is part of human nature to believe in something beyond the grave. Wherever people have been discovered, however basic their living conditions, there is always a belief in a spirit world beyond this one and a conviction that there is life after death. This is because we are created with "eternity in our hearts" (see Ecclesiastes 3:11).

The Christian Bible is not a book of religious sayings jumbled together with no significant order. On the contrary, compiled over more than 1,500 years, it traces a clear progression of God's purpose for the human race from the moment of creation to the end of time. The Bible is, without doubt, the most influential book the world has ever known and it is full of history. A history that has been checked out again and again and found to be reliable.[2]

Throughout the Jewish Scriptures (the Christian Old Testament), the ultimate focus is upon the coming of the Messiah. All the recorded history of Israel, and so much of the preaching of the various prophets, are preparing the way for Jesus Christ who fulfilled detailed prophecies, including the place of His birth, and the manner of His life, teaching, and eventual death.[3]

The ultimate purpose of Jesus Christ was to restore that broken relationship between God and the human race.

2 For a book in this series demonstrating the historical reliability of the Bible see Anderson and Edwards *Evidence for the Bible* (Leominster: Day One Publications, 2014).

3 See for example Micah 5:2; Isaiah 53:1–12; Psalm 22:1–18.

Who was Jesus Christ?

Throughout His three years of public ministry, Jesus lived a perfect life and could challenge His accusers to prove Him guilty of sin (John 8:46; Hebrews 4:15). No one else has ever lived like this. He claimed equality with God and the authority to forgive sins (John 10:30; Luke 5:21), and His unique miracles reinforced this.

Because of His perfect life, He alone had no need to suffer the penalty of death which resulted from the Fall. Instead, by His death He took the guilt, sin, and punishment of all who place their trust in Him, so that they could be forgiven by God (Hebrews 9:28; 1 Peter 2:24).

His death was not the end

Jesus Christ literally and physically rose again from the dead — an event recorded by five independent writers in the New Testament. By this, He proved finally who He was and why He had come. Paul wrote, "Christ died for our sins according to the Scriptures, and that He was buried, and that He rose again the third day according to the Scriptures" (1 Corinthians 15:3–4).

The miracles during His life, His sacrifice on the Cross to bear the guilt of sinners, and His Resurrection from the grave were all part of God preparing the way to restore the whole of the spoiled creation to what it was originally — six days after zero. God has promised that one day Jesus Christ will return to planet Earth and create "new heavens and a new earth, in which righteousness dwells" (2 Peter 3:13). It will be as it was at the beginning, but this time it will never change.

The meaning of life? To enjoy a new relationship of peace with God through the forgiveness of our sins and a new life in friendship with Jesus Christ — and that will bring this broken world closer to what it is intended to be.

COMMENDATIONS

"Be prepared for a roller coaster ride through the world around us — with guides who are leaders in their academic fields. The conclusion is that God's handiwork is everywhere to be found — it is pervasive! This book is an excellent and convincing response to those who claim that a belief in creation requires us to close our minds to evidence.'

David Tyler PhD, MSc, BSc, FTI, MInstP, CPhys, CertEd **Reader at Manchester Metropolitan University**

"Whether we use the microscope or the telescope, whether we consider the living world, the wider universe, or the abstract realm of thought, design that is both wonderful and intricate is clearly seen everywhere. The authors are extremely well qualified in their respective fields to bring these things to our attention. This they have done admirably in this impressive book.'

Prof. Stephen Taylor BSc, MEng, PhD, DSc, DIC, FIEE, FInstPhys, CEng **Professor of Electromagnetics and Physical Electronics, School of Electrical Engineering, Electronics and Computer Science, University of Liverpool**

"When we look carefully at God's creation, it moves us to cry out, 'O LORD, how manifold are thy works! In wisdom hast thou made them all: the earth is full of thy riches' (Psalm 104:24). Here is a beautiful book for adults and young people that displays the marvellous wisdom of God's works from the ant to the Andromeda Galaxy."

Professor Joel R. Beeke, PhD (theology). **President and professor of systematic theology and homiletics at Puritan Reformed Theological Seminary, Grand Rapids, Michigan**

"*Wonders of Creation* is an exhilarating safari exploring the natural world and its setting in the universe. The authors masterfully lay bare the efficiency and aptness of the multitude of specialist designs exhibited by both living organisms and celestial objects. . . . It's sure to fuel more New Atheist anger."

Dr John C. Walton BSc, PhD, DSc, CChem, FRSC, FRSE, **Research Professor of Chemistry, University of St. Andrews, UK.**

"On every page we discover that what we have been led to accept as merely random 'wonders of nature' are in fact the product of deliberate design. Upon the basis of this 'more reasonable explanation' of the physical universe, the whole message of the Bible is seen to fall into place and make perfect sense. Left on the coffee table, I can imagine no more winsomely provocative conversation starter than *Wonders of Creation*.'"

Rev. Jonathan Stephen BA, MSocSc **Principal, Union School of Theology, Wales, UK**

"As the authors were guiding me on this exciting journey of God's wonderful design in His creation . . . my heart was praising 'You are worthy our God to receive glory and honour and power, for you created all things and by your will they were created and have their being.' (Revelation 4:11)."

Dr Nagy Iskander. MB. Bch FRCS (London) FRCS (Glasgow)
Associate Specialist in Surgery (retired)

"*Wonders of Creation* invites us to learn from the natural world, and in particular the remarkable feats of ingenuity which it displays. Only then can we really appreciate how implausible it is for all this to have arisen without the hand of a wise Designer."

Dr Stephen Morris PhD (Physics)
Industrial Physicist

"To read these chapters is to be amazed at the wealth of design in the world around us. The only adequate conclusion to which the evidence points is irresistible: God alone is Creator and is worthy of our obedience, faith and love."

Maurice Roberts BA, BD
Former Lecturer in New Testament Studies in the Free Church of Scotland (Continuing) Seminary.

"This book points us to the evidence for God as Creator, not only in the obvious design features in so many plants and animals, but equally in the physical universe beyond the earth. I recommend it to Christian believers to strengthen faith, to agnostics seeking to answer the great questions of life, and to atheists as an opportunity to reconsider their beliefs!"

Dr Chris Pegington BSc, PhD
Lecturer in genetics at Aberystwyth and Cambridge now a Grace Baptist Minister

"Have you ever wondered how the light from stars so far away reached the earth in such a short time, or why nature has so many examples of regular patterns and shapes? In *Wonders of Creation,* Professors Stuart Burgess and Andy McIntosh answer some of your burning questions in a hugely varied tour of all things created."

Dr Timothy Wells. BSc, PhD
Senior Lecturer in Neuroendocrinology, School of Biosciences, Cardiff University Co-Theme Lead for Endocrinology & Metabolism, The Physiological Society

"A most God-glorifying book! Each page displays the fingerprints of God. It brings us to worship God. How excellent is His name in all the earth! His glory above the heavens. Written by experts in the field, and attractively presented, it makes the evidence for Creation accessible to all."

Dr Matthew J. Hyde BSc (Hons), ARCS, PhD (Biomedical Science), DIC, MClinRes
Research Fellow in Neonatal Medicine at Imperial College London, and Baptist minister

"This book shows that there is a harmony between scientific observation and the words of Scripture. This accord has been missed by naturalistic scientists because they use circular reasoning to reject God's involvement in creation. The fossils, rocks, and the intricate delicate parts of living creatures only point to direct creation, not evolution."

Dr John D Matthews BSc, PhD, CGeol
Licensed Lay Minister – Reader, Church of England.
Retired chartered geologist with over 30 years in the upstream oil industry.
Six years as visiting professor in Earth Science at Imperial College, London.

"Many scientists claim that we cannot find any indication of a Creator in nature. Some even guess the world looks exactly as if there was never a plan behind it. This wonderfully illustrated book will convince you of the contrary — the more scientists explore, the more amazing appears the design in creation. The plethora of evidence argues today more than ever for a Creator."

Dr Reinhard Junker PhD (theol. Int)
Research Assistant
Studiengemeinschaft Wort und Wissen

"This is a remarkable book. In its pages our incredible universe is displayed in its wonder and intricacy. Though committed Christians, including those who are scientists, differ as to how God chose to work, here the sheer brilliance of design and the argument of irreducible complexity are presented with compelling and powerful authority."

Professor Julian Evans OBE, BSc, PhD, DSc, FICFor
Formerly professor of Tropical Forestry, at Imperial College, London.

"*Wonders of Creation* is a masterfully illustrated collection of scientific evidence that confirms biblical creation. From the intricate design found in living creatures, to the stunning beauty of the solar system, all creation reveals the majesty of the Lord."

Dr Jason Lisle Ph.D (astrophysics)
Director of Physical Sciences at the Institute for Creation Research

"Contemporary western culture has widely adopted the assumption that our world is the consequence of natural evolutionary processes which have no ultimate purpose. The authors provide compelling evidence from the world of science and nature of the astonishing complexity pointing to the work of an all-powerful and wise designer God who is greater than we could ever imagine. They present the best contemporary arguments for a literal interpretation of Genesis and in favor of a young earth. Whatever our view of the origin of the universe, their case deserves respect and serious consideration. No one who engages with it will fail to be amazed by the wonders of our world, and they will be challenged to consider afresh the foundational Christian claim that can only be explained by the God of the Bible."

Rev. John Stevens MA, BCL
National Director, Fellowship of Independent Evangelical Churches

"This beautifully illustrated book shows how irreducible complexity is built into every aspect of our universe. Fascinating examples are provided, full of relevant detail, establishing the case for creation as against evolution. Strong arguments clearly expressed and wonderfully supported by pictures and diagrams make this an attractive and informative guide to this vital topic."

Robert Strivens, PhD
Principal, London Seminary

"What a tour de force from two eminent scientists, and a respected pastor-theologian. From stars to sea snails, from galaxies to giraffes, the Creator's designs are detected, and then beautifully, fascinatingly, and compellingly presented. In the words of a well-known hymn, I was moved afresh to acknowledge, How great is God Almighty who has made all things well. I believe you will be similarly affected!"

Dr Steve Brady, Moorlands College, Christchurch, UK

"*Wonders of Creation* describes and illustrates many amazing examples of design in the natural world — from the remarkable traits and abilities of animals to the extraordinary diversity of planets and stars and the exquisite design of the human body. The message of the authors is clear: all display the glory, beauty, and majesty of the God who made them."

Paul Garner BSc, MSc, FGS
Masters in Geoscience (specializing in palaeobiology)
Researcher and Lecturer with Biblical Creation Trust

Daily Lesson Plan

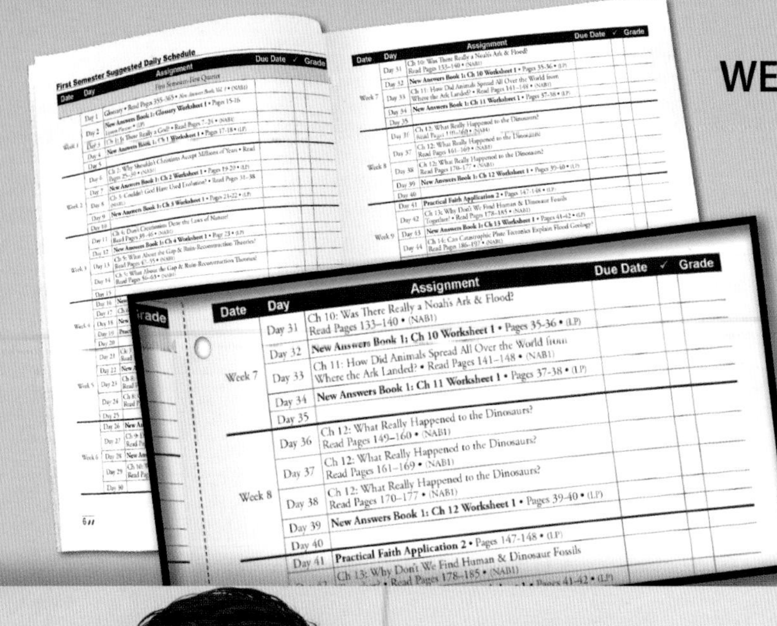

Date	Day	Assignment	Due Date	✓	Grade
Week 7	Day 31	Ch 10: Was There Really a Noah's Ark & Flood? Read Pages 133–140 • (NAB1)			
	Day 32	**New Answers Book 1: Ch 10 Worksheet 1** • Pages 35-36 • (LP)			
	Day 33	Ch 11: How Did Animals Spread All Over the World from Where the Ark Landed? • Read Pages 141–148 • (NAB1)			
	Day 34	**New Answers Book 1: Ch 11 Worksheet 1** • Pages 37-38 • (LP)			
	Day 35				
Week 8	Day 36	Ch 12: What Really Happened to the Dinosaurs? Read Pages 149–160 • (NAB1)			
	Day 37	Ch 12: What Really Happened to the Dinosaurs? Read Pages 161–169 • (NAB1)			
	Day 38	Ch 12: What Really Happened to the Dinosaurs? Read Pages 170–177 • (NAB1)			
	Day 39	**New Answers Book 1: Ch 12 Worksheet 1** • Pages 39-40 • (LP)			
	Day 40				
	Day 41	**Practical Faith Application 2** • Pages 147-148 • (LP)			

WE'VE DONE THE WORK FOR YOU!

PERFORATED & 3-HOLE PUNCHED
FLEXIBLE 180-DAY SCHEDULE
DAILY LIST OF ACTIVITIES
RECORD KEEPING

"THE TEACHER GUIDE MAKES THINGS SO MUCH EASIER AND TAKES THE GUESS WORK OUT OF IT FOR ME."

☆☆☆☆☆

Master Books® Homeschool Curriculum

Faith-Building Books & Resources
Parent-Friendly Lesson Plans
Biblically-Based Worldview
Affordably Priced

Master Books® is the leading publisher of books and resources based upon a Biblical worldview that points to God as our Creator.